Risk/Benefit Analysis in Water Resources Planning and Management

Risk/Benefit Analysis in Water Resources Planning and Management

Edited by Yacov Y. Haimes

Center for Large Scale Systems and Policy Analysis
Case Western Reserve University
Cleveland, Ohio

PLENUM PRESS • NEW YORK AND LONDON

Library of Congress Cataloging in Publication Data

Engineering Foundation Conference on Risk/Benefit Analysis in Water Resources
  Planning and Management (1980 : Pacific Grove, Calif.)
  Risk/benefit analysis in water resources planning and management.

  Bibliography: p.
  Includes index.
    1. Water resources development — Congresses. 2. Risk — Congresses. 3. Decision-
  making — Congresses. 4. Technology assessment — Congresses. I. Haimes, Yacov Y. II.
  Universities Council on Water Resources. III. Title.
  TC401.E53   1980   333.91′15   81-17824
  ISBN 0-306-40884-8   AACR2

Proceedings of an UCOWR-sponsored Engineering Foundation Conference
on Risk/Benefit Analysis in Water Resources Planning and Management
held at the Asilomar Conference Grounds, Pacific Grove, California,
September 21-26, 1980

© 1981 Plenum Press, New York
A Division of Plenum Publishing Corporation
233 Spring Street, New York, N.Y. 10013

FOREWORD

Ronald M. North

President
Universities Council on Water Resources

People sense intuitively that the world in which we live is not free of risk. Every decision, every action, even the refusal to either act or decide involves some element of risk. Perhaps, because we accept relatively low levels of risk in our daily activities, we tend to minimize the existence of risk and thereby fail to include risk assessment in those decisions and actions which could be improved through a risk assessment process. However, our casual approach to risk assessment seems to stem largely from the difficulties inherent in measuring risk rather than from any lack of cognizance of the existence of risk. This conclusion is evidenced by the many statements in official documents relating to planning and evaluation which suggest that risk assessments should be conducted but do not provide the mechanism for such assessments nor do they encourage their consideration in the decision making process.

This conference on Risk/Benefit Analysis in Water Resources Planning and Management is notable because it attempts to identify and evaluate the mechanisms available for risk assessment which might be useful in water resources planning and management efforts. These proceedings bring together the thoughts of professional persons who have struggled with the problems of risk assessment and who have contributed to the refinement of both theoretical and pragmatic solutions for the improvement of risk assessment processes. However, the most significant aspect of these proceedings is that they reflect the opportunities for the exchange of ideas among professionals of various disciplines and among those who are involved with various approaches in different applications. I believe this conference makes a significant contribution to advancing the state-of-the-art as it is now developing in risk/benefit analysis and it

v

certainly has enhanced the opportunities for further research and
development among those who participated in the discussions at the
Asilomar Conference Grounds. Those who have attended such conferences at Asilomar will appreciate the unsurpassed opportunities for
the interaction of minds and the power of those interactions as they
are focused on a specific topic without the distractions of the outside world. Hopefully, these proceedings will reflect the advance-
ments in the state-of-the-art for those who were unable to participate directly.

The comments of professionals involved in risk assessment owe
a debt of gratitude to Yacov Haimes for his initiation, funding,
implementation, and completion of this conference and these proceedings. Without his single-minded dedication to the issues he so
clearly recognizes in risk assessment, the conference and these
proceedings would not have happened. The Universities Council on
Water Resources was pleased to sponsor the resolution which initiated this conference and we were very happy to join the U.S. Water
Resources Council and the Office of Water Research and Technology
in cosponsoring the conference so ably hosted by the Engineering
Foundation.

PREFACE

The UCOWR-sponsored Engineering Foundation Conference on Risk/
Benefit Analysis in Water Resources Planning and Management was held
at the Asilomar Conference Grounds, Pacific Grove, California, September 21-26, 1980, and was attended by about sixty participants
from academia, industry, Congress, and federal, state, and local
government.

The conference's goals and objectives were to:

1) Familiarize the participants with the state-of-the-art in
   risk/benefit analysis.

2) Explore the feasibility of using risk/benefit analysis in
   water resources planning and management.

3) Provide a medium conducive to the exchange of information
   on the conference theme among educators, analysts, managers,
   and policymakers.

4) Identify and articulate future desired actions designed to
   alleviate some of the present problems we face in risk/
   benefit analysis and risk assessment in general.

These proceedings attempt to provide some answers to myriad
questions of risks and uncertainties that arise in the formation of
public policy in general, and in water resources planning and
management in particular. A sample of some generic questions is
listed below:

a) What is the efficacy of models on risk analysis and decision
   making?

b) To what extent are these models and methodologies credible?

c) To what degree does the value of information increase with
   more models on risk?

d) How important are the methodologies for the quantification
   of risk as part of the risk assessment process, and how
   critical is the understanding of the process itself?

e) Who should decide on acceptability of what risks, for whom,
   and in what terms, and why? [W. W. Lowrance]

I would like to thank the Universities Council on Water Resources, the Engineering Foundation, the U.S. Water Resources Council, the Office of Water Research and Technology, U.S. Department
of Interior, the Conference Steering Committee, the conference
speakers and participants, and Plenum Press for their support,
assistance, cooperation, and encouragements.

I particularily would like to thank the Editorial Staff of
the Center for Large Scale Systems and Policy Analysis, Case
Western Reserve University--Virginia Benade, Joyce Martin, and
Mary Ann Pelot--for their outstanding help.

Yacov Y. Haimes

CONTENTS

SESSION I: TRADE-OFFS IN RISK/BENEFIT ANALYSIS
Chairman: Yacov Y. Haimes
Case Western Reserve University

Opening Remarks and Conference Overview   1
Yacov Y. Haimes, Conference Chairman

Uncertainty: The Water Resources Decision
Making Dilemma   5
Leo M. Eisel, Water Resources Council (formerly)

Technological Hazards, Risk, and Society: A Perspective on Risk Analysis Research   13
Vincent T. Covello, National Science Foundation

Risk/Benefit Trade-off Analysis in Water
Resources Planning   31
Warren A. Hall, Colorado State University

SESSION II: IDENTIFICATION AND MEASUREMENT
OF RISK PARAMETERS AND OBJECTIVES
Chairman: Vincent T. Covello
National Science Foundation

Minimizing Risk of Flood Loss in the National
Flood Insurance Program   41
Malcolm Simmons, The Library of Congress

What Kind of Water Will Our Children Drink?   53
David Okrent, University of California

SESSION III: METHODOLOGIES IN (a) MEASURING AND QUANTIFYING
RISK AND (b) RISK-BENEFIT TRADE-OFF ANALYSIS
Chairman: Warren A. Hall
Colorado State University

Methodology and Myth   59
William D. Rowe, The American University

Risk-Benefit Analysis in a Multiobjective Framework   89
Yacov Y. Haimes, Case Western Reserve University

Multiobjective Generating Techniques for
Risk/Benefit Analysis   123
Jared L. Cohon and Charles S. ReVelle,
The John Hopkins University
Richard N. Palmer, University of Washington

The Risks of Benefit-Cost-Risk Analysis   135
Ronald A. Howard, Stanford University

Methods for Determining the Value of Model Development in Cost/Benefit/Risk Analysis   149
M. Scott Nainis, Arthur D. Little, Inc.

SESSION IV: IMPLEMENTING THE RISK AND UNCERTAINTY PROVISIONS
OF THE PRINCIPLES AND STANDARDS
Chairman: Gary Cobb
U.S. Department of the Interior

Application of Risk and Uncertainty Analysis in the
Principles, Standards, and Procedures of the
U.S. Water Resources Council   157
David C. Campbell, U.S. Water Resources Council

Risk Analysis Applicable to Water Resources Program
and Project Planning and Evaluation   163
Ronald M. North, The University of Georgia

Risk Assessment from a Congressional Perspective   175
James W. Spensley, U.S. House of Representatives

Risk Assessment: The Role of Government in a
Multiple Objective Framework   181
Warren A. Hall, Colorado State University

Implementing the Risk and Uncertainty Provisions
        of the Principles and Standards   191
    David C.N. Robb, St. Lawrence Seaway Development
    Corporation

        SESSION V: SOCIETAL PREFERENCES FOR RISK AND SAFETY
                     ASSESSMENT AND EVALUATION
                     Chairman: James W. Spensley
                     U.S. House of Representatives

Rating the Risks   193
    Paul Slovic, Baruch Fischhoff, and Sarah
    Lichtenstein, Decision Research

Risk Assessment: Arid and Semiarid Lands Perspective   219
    J. Eleonora Sabadell, U.S. Department of the Interior

The Value of a Life: What Difference Does It Make?   233
    John D. Graham, National Academy of Sciences
    James W. Vaupel, Duke University

On the Value Dependent Role of the Identification
        Processing and Evaluation of Information in
        Risk/Benefit Analysis   245
    Andrew P. Sage and Elbert B. White, University of
    Virginia

                SESSION VI: PANEL DISCUSSION
                Chairman: Jonathan W. Bulkley
                     University of Michigan

The Role of Government in Assessing the Acceptability
        of Risk and the Efficacy of Safety   263
    Jonathan W. Bulkley, University of Michigan (Chairman)
    Theodore M. Schad, National Research Council
    David S. Bowles, Utah State University
    J. Eleonora Sabadell, U.S. Department of the Interior
    Malcolm Simmons, The Library of Congress

Summary, Conclusions, and Recommendations   277

List of Participants   283

Index   287

OPENING REMARKS AND CONFERENCE OVERVIEW

         Yacov Y. Haimes

         Conference Chairman

     On behalf of the conference sponsors--the Universities Council
on Water Resources (UCOWR), the Water Resources Council (WRC), and
the Office of Water Research and Technology (OWRT), U.S. Department
of Commerce--I would like to welcome you and extend my appreciation
to the Engineering Foundation for organizing the conference. Special
thanks are due to Drs. Sanford Cole and Harold Comerer of the Engineering Foundation, Dr. Leo Eisel, Director of WRC, Mr. Gary Cobb,
Director of OWRT, and Dr. Ronald North, President of UCOWR. Finally
I would like to thank the Conference Steering Committee whose efforts
made the success of this meeting possible; they include: Jonathan
Bulkley, Gary Cobb, Leo Eisel, Robert Friedman, John Graham, Warren A.
Hall, Steve Hanke, W. Scott Nainis, Herbert Snyder, James (Skip)
Spensley, James Vaupel, and Warren Viessman, Jr.

     At its annual business meeting on August 1, 1979, the Univer-
sities Council on Water Resources adopted a resolution
requesting the Engineering Foundation to conduct a conference on
"Risk/Benefit Analysis in Water Resources Planning and Management."
Central to this resolution was the recognition that risk elements
characterize almost all water resources activities, that risk/benefit
analysis is still in its infancy as a methodological procedure, yet
it is being proposed as a procedure for assessing, evaluating, and
comparing alternative policies, and that the water resources professional community should take a lead in contributing to the
advancement of the state-of-the-art of this important area.

     This conference is being convened at a time when major environ-
mental risk and health safety problems associated with water resources
are facing the nation. Although water experts have long recognized
that most, if not all, water resources projects are planned, designed,

constructed, operated, and managed under risk and uncertainty, it is only recently that public awareness and concern regarding environmental risk and safety have reached an epidemic level. Among the risk-related water resources problems on the nation's agenda are:

- contamination of aquifers due to waste disposal and other man-made activities

- contamination of surface water resources due to controlled and uncontrolled point and nonpoint sources of pollution

- drinking water problems associated with water treatment by such chemicals as trihalomethenes

- acid rain due to the burning of fossil fuel

- structural hazards and dam safety

- floods

- droughts and other water shortages

- desertification

- endangered species

- etc.

The above representative list of lingering issues deserves to be studied and evaluated by compatible risk/benefit methodologies. One of the major objectives of this conference is to ascertain the extent to which the respective risk/benefit methodologies are indeed available and credible. Although the field of risk/benefit analysis is still in its infancy--representing about a decade of recent concerted efforts since J. von Neumann and O. Morgenstern's 1947 book, M. Friedman and L. Savage's 1948 paper, and K. Arrow's 1951 paper-- a welter of literature already exists. Prominent among these contributions are Chauncy Starr's paper in Science in 1969, the National Academy of Engineering COPEP report in 1973, William Lowrance's 1976 book "Of Acceptable Risk," William Rowe's 1977 book "Anatomy of Risk," the works of Paul Slovic and his group, Howard Raiffa and his group, etc.

Risk/benefit analysis associated with water resources problems transcends multifarious aspects including science and technology, human well-being, societal norms, traditions, philosophies of life and life styles, institutions, laws, and politics. To adequately and credibly respond to these aspects, risk/benefit analysis must be holistic in nature. Imperatively, risk/benefit analysis--which should be responsive to different problems at different times, affecting different constituents in different locations and caused by different agents--will require different treatments, methodologies, approaches, and data bases. Obviously, no one methodology of or approach to risk/benefit analysis--powerful and effective as it might be--would be adequate in addressing the myriad problems associated with risk.

Probably the most critical as well as controversial aspect associated with risk/benefit analysis is the acceptability of risk. William Lowrance articulated this problem in one fundamental question: "Who should decide on acceptability of what risks for whom, and in what terms, and why?" We may never be able to answer these questions generically. All we can hope for are specific, often fuzzy, answers to very specific issues for specific time references and under certain circumstances and specific societal, political, and institutional conditions.

This assembly of sixty professionals, although it represents a diversity of background, is unified in its objective towards improving its knowledge in risk/benefit analysis in water resources planning and management. The Conference Steering Committee has purposely invited to the program several experts in risk/benefit analysis who might not necessarily be experts in the water resources field. Their participation in this meeting will ensure a cross fertilization among the various disciplines. I am sure that I speak for all of us when I say that the exclusive scenic beauty of Asilomar and Monterey Bay coupled with the very promising scheduled technical program offer an exciting week for all of us.

UNCERTAINTY: THE WATER RESOURCES DECISIONMAKING DILEMMA

Leo M. Eisel

U.S. Water Resources Council
Washington, D.C.

I appear before you today as an ex-bureaucrat. Consequently,
I am able to tell you about risk and uncertainty and how it affects,
or does not affect, decisionmaking in the Federal Government. I
can finally tell it like it really is--without thoughts of loyalty
to my former employer, the Federal Government. I plan to lay it
all out and be the Jack Anderson on risk and uncertainty in Federal
decisionmaking.

First, however, I would like to go back to a period of about
ten to twelve years when I was pursuing the lowliest of all tasks--
being a graduate student and writing a Ph.D. thesis. The purpose
of this twelve-year flashback is to convince you that I am an old
hand on risk and uncertainty in decisionmaking and am simply not
here to tell a few war stories about my frustrations in insuring
more adequate treatment of risk and uncertainty in Federal decisionmaking. In the late 1960s I was working on a thesis topic
which concerned development of a chance-constrained programming
model of a land and water system. For the few of you here who do
not know what chance-constrained programming is, it is simply a
procedure which allows you to incorporate random variables into
mathematical programming models. In order to specify chance-
constrained programming models, you must have some estimate of the
statistical characteristics of these random variables, but do not
necessarily have to know their distributions. This particular
chance-constrained programming model incorporated the random variation from streamflow as well as the variance of other parameters--
for example, reservoir sedimentation in an attempt to optimize
reservoir design.

Over the past ten years or so, I have on a few occasions found my thesis work quoted in various journals, but am unaware of any attempts to employ this chance-constrained programming model in any real world decisionmaking. Furthermore, it is interesting to look at how my particular work has been quoted.

Generally, it is used as an example of the difficulties of using relatively sophisticated methods, such as chance-constrained programming, to incorporate risk into decisionmaking. These difficulties originate because of the mathematical complexities in solving the model and the need for considerable data in order to estimate the statistical properties of the random variables.

My purpose for telling you this little story concerning my early experience with risk and uncertainty is to demonstrate that in some cases at least the world is not ready for relatively sophisticated methods to incorporate risk and uncertainty into decisionmaking.

Now, while keeping this little story about the fate of my chance-constrained programming model in the back of your mind, let us move down the road approximately eight years until July of 1978. This was an important month to me since on July 12, 1978, President Carter's water policy directives--which resulted from more than a year's work of scrutiny of Federal water resources policy and programs--were announced. One of these directives from two years ago went to the Water Resources Council and ordered the Council to carry out a thorough evaluation of agency practices for making benefit and cost calculations and to publish a planning manual that would ensure that benefits and costs are estimated using the best current techniques and calculated accurately, consistently, and in compliance with the Principles and Standards and other appropriate economic evaluation requirements. Specifically, the President stated that the new benefit cost procedures should, among other things, pay special attention to improved methods for considering the uncertainty and risk of costs and benefits. Furthermore, the President directed that attention should be given to "consideration and display of engineering uncertainty," that is: risk and uncertainty of structure failure. Therefore, here it was July 1978 and the President had just directed the Water Resources Council to develop some rules and regulations for use by the Federal agencies to more adequately consider risk and uncertainty in water resources decisionmaking. In addition, the President had specifically highlighted the need to adequately incorporate consideration of structural failure into benefit-cost calculations.

The general vehicle for carrying out these Presidential directives was the Principles and Standards for Planning Water and Related Land Resources. For those of you who are not familiar with water resources planning, the Principles and Standards are a set of rules and regulations published in the Federal Register by the Water Resources Council which form the basis of planning and analysis procedures used by the Corps of Engineers, the Water and Power Resources Service, the Soil Conservation Service, and the Tennessee Valley Authority in the planning and design of Federal water resources projects. These Principles and Standards provide regulations for planning and designing approximately three billion dollars worth of Federal water resources construction per year, and are the basic rules for benefit-cost calculations.

Now that I have set the stage, I can get to the real meat of my Jack Anderson-style exposé of the treatment of risk and uncertainty in Federal water resources decisionmaking. I'd like to take a bit of your time this morning to tell you about what happened over the past two years in the Water Resources Council's efforts to develop a set of rules and regulations which could be used by Federal water resources planners to carry out the Presidential directives I've just described. I think it is instructive to look at what happened over the past two years in the very nitty-gritty world of water resources planning and then compare it to my ivory tower experience of about ten years ago. I think that looking at the experience of the Water Resources Council over the past two years will provide instructive guidance to the conference over the next week.

Basically, the Council interpreted the Presidential directive to mean that two basic areas of risk and uncertainty should be addressed in the Principles and Standards. First, procedures needed to be developed for adequately incorporated consideration of structural failure into water resources planning. In 1977 and 1978, when much of the work on the President's water policy review was done, the memory of the Teton Dam failure in 1976 was fresh in everyone's mind; consequently, the directive concerning engineering risk and uncertainty was specifically designed to have the Council develop a procedure that would adequately incorporate the costs and benefits of a structural failure like the Teton Dam, into Federal water resources decisionmaking. What took a portion of one sentence in the Presidential directive was soon translated into almost two years of work with countless meetings, high-powered outside consultants, people drawing Poisson distributions on blackboards, and a lot of very interesting discussion at the Council concerning how to translate statistical theory into something which could be reasonably, reliably and consistently implemented by, for example, a district office of the Corps of Engineers. The statistical theory for incorporating the risk of structural failure, such as the Teton Dam failure, into water resources decisionmaking is very simple--you calculate the expected value of the costs of structural failure and incorporate

that into benefit-cost analysis. The major problem which developed, however, was estimating the probability of structural failure. While there have been numerous failures of water resources
structures in the United States, only a relative handful--less than
about a dozen--have occurred in this century on dams which were
well-engineered and constructed according to modern practices. The
majority of dam failures have occurred on small dams, which were
poorly engineered and constructed with now outmoded methods. Furthermore, the well-engineered structures that have failed have been
of a large variety of types and designs. After a considerable
period of time, lots of outside consultants and innumerable meetings, it became clear that the practical problems of estimating
probabilities of structural failure of dams were very great. In
short, the sample of structural failures for various types of dams
was simply too small to allow estimation of statistical parameters
with any kind of reasonable certainty.

Furthermore, there proved to be significant political problems
in developing procedures for incorporating probability of structural failure into Federal water resources decisionmaking which
would be acceptable to the Federal agencies. This is not necessarily a criticism of the Federal agencies, since an agency such
as the Corps of Engineers produces a product--a dam--and they are
understandably reluctant to accept a procedure which requires them
to estimate the probability of failure of what is almost certainly
a well-engineered and dependable structure. This is not a characteristic unique to Federal water resources agencies. My point,
however, is that the problems with developing procedures for consideration of risk and uncertainty in decisionmaking are not all
technical. There are very real political and legal problems in
acknowledging the probability of failure of your product--whether
it is a dam, a Ford Pinto, or a nuclear power plant. These legal
and political problems are such that, unless the data and theory
supporting these procedures are absolutely solid, it may be very
difficult, or impossible, to actually gain widespread acceptance
of the procedures regardless of the validity of the need to incorporate risk and uncertainty into decisionmaking.

After two years of effort, a simplified procedure for incorporating risk and uncertainty of structural uncertainty into Federal water resources decisionmaking was adopted. This procedure
was approved by the Council at its September 1980 meeting and will
be published shortly in the Federal Register as a final rule.
Basically, this simplified procedure does not require estimation
of dam or other structural failure probabilities. It merely requires that a map be produced and included with the other project
planning and design documents. This map should show the aerial
extent of flooding which would be caused by structural failure,
the number of people living or working in this area, and finally,

the value and number of properties which would be adversely affected by flooding caused by structural failure.

I support this final decision because I think that adequate data do not exist to obtain reasonably good estimates of structural failure. However, I have told you this little story in an effort to demonstrate that many obstacles confront even simple measures to incorporate risk and uncertainty in water resources decisionmaking.

The second portion of the July 1978 Presidential directive dealt with risk and uncertainty in benefit-cost calculations. In benefit-cost analysis of proposed water resource projections, the sources of risk and uncertainty are: (1) the variability of complex natural, social, and economic situations, and (2) the fact that future demographic, economic, hydrologic and meteorologic events are essentially unpredictable because they are subject to random influences. Again, a significant amount of effort over the past two years has been spent on developing rules and regulations for incorporation of this form of risk and uncertainty into water resources decisionmaking at the Federal level. The end product of this work is described in the Friday, December 14, 1979, Federal Register in the section dealing with the procedures for implementing Principles and Standards and will be further detailed in the final rules and regulations to be published this fall in the Federal Register.

In the published rules and regulations, it has been left to the planner to characterize to the extent possible the different degrees of risk and uncertainty and describe them clearly so that decisions can be based on the best available information. The planner should also suggest adjustments in design to reflect the various attitudes of decisionmakers toward risk and uncertainty. Furthermore, the rules state that if the planner can identify in qualitative terms, the uncertainty inherent in important design, economic, and environmental variables, these judgments can be transformed into a subjective probability distribution.

The planner is encouraged, where appropriate, to characterize the range of outcomes with a set of subjective probability estimates, but the project report must make clear that the numerical estimates are subjective. Moreover, subjective probability distributions must be chosen and justified case by case and some description of the impact on design and using other subjective distributions must be given. The rules also permit sensitivity analysis--that is, testing the sensitivity of the outcome of project evaluation to variation of the magnitude of key parameters.

I hope that it is clear that the rules and regulations give

a great deal of latitude to the planners in the field doing the
work. Where the data exist and the need exists, fairly sophisticated Bayesian models can be utilized. However, no attempt was
made to require the use of Bayesian statistics or other subjective
statistical methods, for incorporating consideration of risk and
uncertainty into the water resources planning and design.

While one can be critical of the latitude in the published
rules that is left to the individual planner--whether he or she is
a planner in the Rock Island District of the Corps, or the California region of the Water and Power Resources Service--The Council
concluded that it was not possible to adopt across the board blanket requirements for sophisticated procedure to incorporate risk
and uncertainty into decisionmaking.

Furthermore, we gave a lot of thought to what would be useful
to the decisionmakers--that nebulous combination of the President,
Congress, local conservancy districts, local project sponsors,
mayors, etc. For example, a simple statistical parameter, such as
standard deviation, is a key element in the quantitative incorporation of risk and uncertainty into the decisionmaking. Can, however,
we assume that these decisionmakers in general will understand and
utilize the concept of standard deviation? I don't think we can
necessarily assume this. I think there is a great problem in providing more sophisticated information concerning risk and uncertainty than can ever be effectively used. Consequently, I again
support the general procedures approved and published by the WRC
for incorporating risk and uncertainty into Federal water resources
decisionmaking. Granted, there is potential for abuse and for not
going the extra mile in many cases and putting together adequate
models for incorporating risk and uncertainty. Yet, on the other
hand, I believe the final rules and regulations approved by the
Council avoid the dangers of a blanket requirement for sophisticated procedures when the data are not there or when the computational ability of Federal agency field offices is not adequate.
Furthermore the rules do not require the generation of excessively
sophisticated information which cannot be utilized by decision-
makers.

Well, where are we? I have tried to give you a range this
morning of the possibilities for incorporating risk and uncertainty into water resources decisionmaking. On the one hand, we
have the sophisticated procedures, such as my little experience
with chance-constrained programming, which can produce good theoretical models for incorporating risk and uncertainty into decisionmaking. On the other hand, I've tried to give you a few war
stories concerning the very real problem caused by lack of data,
lack of computational capability, and political objections which
can create obstacles to consideration of risk and uncertainty in

water resources decisionmaking at the Federal level.

Over the next five days you are going to spend a lot of time thinking about better ways to incorporate risk and uncertainty into water resources and other areas of decisionmaking. I would like to leave a couple of thoughts with you. You may not all agree with these thoughts. However, I believe that part of my purpose in this speech this morning should be to stir things up a bit. From my experiences, both at Harvard University as a graduate student writing about chance-constrained programming and as Director of the Water Resources Council in carrying out a Presidential directive specifically designed to incorporate risk and uncertainty into water resources decisionmaking--I believe, in many cases, that the real challenge and need is not for developing bigger, better, more sophisticated chance-constrained programming models or Bayesian statistical models, but rather with sitting down and developing very simple-minded procedures. These procedures must be such that data, computational ability and capability and willingness by decisionmakers to use the output all exist. I am not very sure in the next twenty-five years that we are going to have many decisionmakers who really understand the concept of a standard deviation, let alone a Bayesian statistical model. This is not a criticism but simply the fact that they are not statisticians. A basic problem, however, is that pursuit of these kind of simple models would not have gotten me a Ph.D. at Harvard University, while the chance-constrained programming model did. I am fully aware of this. Furthermore, I am aware of the need to continue to probe the outer limits of our knowledge concerning risk and uncertainty and the incorporation of this into decisionmaking procedures; I am not arguing against basic or theoretical research. Rather, I am simply stating that if we are gathered here this week to speak to major concerns of incorporating risk and uncertainty into decisionmaking, that we should spend at least part of our time talking about some very simple-minded models for incorporating consideration of risk and uncertainty into real world situations. I realize that some of you come from areas of expertise outside of water resources and much of what I have said here today may not strictly apply to you. You may have decisionmakers in your fields who more fully understand the concept of a standard deviation and may be able to use more sophisticated models. However, I do ask that you think about the need for simple-minded procedures and not be captivated solely by more and more sophisticated models.

TECHNOLOGICAL HAZARDS, RISK, AND SOCIETY: A PERSPECTIVE ON

RISK ANALYSIS RESEARCH*

Vincent T. Covello

Division of Policy Research and Analysis
National Science Foundation
Washington, D.C.

INTRODUCTION

In recent years government support for the study of technological hazards and risks has substantially increased. Much of the
research effort has been carried out through grants and contracts
to statisticians, physicists, biologists, chemists, engineers, decision analysts, economists, and psychologists, although researchers
from other disciplines have also made major contributions. Largely
due to these efforts, there is now a significant body of knowledge
on methods for identifying estimating, and evaluating risks; on
institutional, political, and legal constraints associated with the
risk management process; and on public perceptions of risk. It
would be no exaggeration to say that as a result of these efforts
a new and lively field of research has been created almost overnight.

Despite this progress, there are still major gaps in our knowledge. Not only has the research effort raised new questions, but
recent events have focused attention on a new set of issues. For
the most part, these issues concern (1) competition between the
goals of risk reduction and other national objectives, such as reducing inflation and increasing economic productivity; and (2) disagreements and differences among scientific experts, policymakers,
and the lay public in their risk estimates and perceptions.

----

*Note: The views expressed in this paper do not necessarily repre-
sent the views of the National Science Foundation, but are exclu-
sively those of the author.

Although researchers are beginning to address these issues
directly, relatively little time will be spent in this paper reviewing these efforts or the more general risk analysis literature.
For such a review, the interested reader is referred to a recent
overview article on issues in risk analysis [Covello and Menkes,
1981] and to the bibliography on risk analysis and risk management
appended to this paper [Appendix A]. The principal objective of
the present paper is, instead, to discuss the use of risk analysis
research in the policy decision-making process.

THE RELEVANCE OF RISK ANALYSIS RESEARCH

Advances in science and technology, as recent events have
demonstrated, are often accompanied or followed by risks to health,
safety, or the environment. Various incidents--such as the Three
Mile Island accident, the grounding of the DC-10's, the Love Canal
controversy, and the almost daily discovery of new hazards--have
focused public attention on the need to anticipate, prevent or
reduce the risks inherent in modern society. As Aaron Wildavsky
[1979] has observed:

> How extraordinary. The richest, longest-lived, best-
> protected, most resourceful civilization, with the
> highest degree of insight into its own technology,
> is on its way to becoming the most frightened. Has
> there ever been, one wonders, a society that produced
> more uncertainty more often about every day life?

Faced with this situation, public and private sector decision
makers are confronted with numerous risk-management choices for
which relevant information is only partially in hand. Information
is urgently needed that would help decision makers and policy
analysts in the following areas:

1. Determine the true range of effects of a large variety
   of technological hazards.

2. Refine the methods for detecting and measuring these
   effects.

3. Assess the economic and social impact of alternative
   risk-mitigation strategies.

4. Establish priorities in the allocation of funds for
   risk management programs.

5. Establish priorities in the determination of safety
   standards.

6. Anticipate demands for information from concerned public groups.

7. Evaluate the stability and credibility of expert and public opinion related to safety issues.

8. Forecast public reactions to safety standards and risk abatement guidelines.

PERSPECTIVES ON THE ROLE OF RISK ANALYSIS RESEARCH IN POLICY DECISION MAKING

By addressing these topics, researchers can provide information that will help decision makers anticipate and plan for technological risks to health, safety, and the environment. It would be naive, however, to expect that such research will have a direct effect on public policy decision making. As recent studies [Wildavsky, 1979] have shown, the link between research and policy decision making is complex and varied. The once-popular paradigm of (a) the appearance of a problem, (b) the identification of knowledge gaps, (c) the commissioning and execution of needed research, (d) the development of policy alternatives and their consequences, and (e) the choice of a preferred solution is no longer accepted as an adequate description of reality. A number of studies in which policy officials were asked about their use of research findings [Larsen, 1980; Weiss, 1977a, 1977b, and 1978; and Lindblom and Cohen, 1978] indicate that this linear paradigm can be seen at work only in rare and usually highly technical instances. Policymakers frequently cite sources such as newspaper and magazine articles for changes in their views of problems, suggesting that informal channels may be an important source of information; and they make use of research in ways far more subtle and complex than the simplest problem-research-solution model would suggest.

New knowledge may precede and, in part, cause the perception of a problem and then become part of the basis for a solution; in an interactive, problem-fraught situation, individuals or groups from various backgrounds may come together, each contributing some knowledge in arriving at a new direction for action. Or, research may be used as ammunition to support positions arrived at on other grounds. Research may be used as a tactic to delay a decision, to show that something is being done about a problem, to enhance the prestige of an agency, to satisfy a constituency, to build support for a program, or as a way of parrying unwelcome demands for taking action. And research may simply be part of much broader cultural currents, along with literature, criticism, history, and journalism, often funded after rather than before policy initiatives have been taken.

Perhaps the dominant interpretation to emerge is that of research as "enlightenment." No single research project's conclusions necessarily influence particular decisions, but over time a cumulative body of research brings new generalizations and concepts that gradually become current in decision-making processes. The steps in this process may be hard to trace, but policymakers slowly change the way they look at issues or define problems and also the way they ask questions. The enlightenment function that research serves is not so much that it provides solutions to problems but that it contributes to changes in the intellectual framework within which problem-solving takes place. Research is thus one of the influences that, over long time periods, alters the terms of the policy debate, shifts the salience of issues, revises the policy agenda, provides new concepts for dealing with old issues, and creates new ranges of possible action.

If the decision-making process is looked at more broadly, it is clear that policy is made by a large number of interacting individuals, each needing specialized information [Knorr, 1977; and Lynn et al., 1975]. Systematic research can never be more than a supplementary source of that information. Often knowledge is less determinative of policy outcomes than attitudes and values, and it is the latter that determine whether new knowledge will be used. Moreover, problems exist for which there is no corresponding problem-solving activity, in fact for which there are no solutions. In these cases, while systematic research may be done, it is unlikely to be of direct use.

These observations have several implications for risk and analysis research. First, researchers, supporters of risk research (see Appendix B), and decision makers need to adopt a realistic perspective on the policy usefulness of risk analysis research. Clear connections between particular studies and specific outcomes will be rare and hard to identify; more impressionistic evidence of subtle shifts in the terms of policy debate may be the most that can be found.

Second, greater effort should be devoted to studying the role of risk analysis research in the decision-making process. Research is far more likely to be useful if it is designed with an awareness of the informational needs of decision makers. Yet these needs cannot be known without careful research on the risk management decision-making process and on the role of research in that process.

Third, since there are a variety of channels through which research enters the policy process, agencies supporting risk research should actively encourage researchers to employ multiple dissemination strategies. Researchers who have been most successful in bringing their research to the attention of decision makers report that they not only submit contract reports but also consult both formally and informally with executive branch and congressional policy officials, give expert testimony in court proceedings, accept temporary appointments in government policy offices, hold news conferences announcing their findings, attend and present papers at a broad spectrum of conferences, sponsor seminars and workshops attended by academic researchers, government policy analysts, and risk management decision makers, invite government policy analysts and risk management decision makers to serve on project advisory committees, and publish their research results in academic journals and popular magazines.

Fourth, the literature on research utilization points to the importance of the within-government research broker. Government agencies with risk analysis or risk management responsibilities should have at least one office specializing in maintaining contact with the research community. The functions of such an office should be to monitor risk analysis research, translate results into the language of the decision makers, analyze alternatives, distinguish between what is known and what is believed, and channel the information to appropriate program offices. In order to perform these functions effectively, brokerage offices should be free of program commitments and should be provided with an extramural research budget.

Finally, it needs to be recognized that risk is in large part a management problem--management on the local level or at the national level. To cope better we need to understand better the nature of risks and how they develop, and to translate this understanding into effective policies. It would be unreasonable, however, to pretend that such understanding can carry with it the certainties and completeness of the physical sciences. In analyzing risks, the central need is to evaluate, to order, and to structure inevitably incomplete and conflicting knowledge so that management choices can be made with the best possible understanding of current knowledge and its limitations.

APPENDIX A: BIBLIOGRAPHY ON RISK ANALYSIS

Arrow, K. J., 1974, "Essays in the Theory of Risk Bearing," Elsevier, New York.
Baram, Michael S., 1980, Cost-benefit analysis: an inadequate basis for health, safety, and environmental regulatory decision-making, Ecology Law Quarterly, 8:473-531.
Baram, Michael S., 1979, "Regulation of Health, Safety and Environmental Quality and the Use of Cost-Benefit Analysis," Final Report to the Administrative Conference of the United States, Washington, D.C.: Administrative Conference of the U.S.

Bazelon, D. L., 1979, Risk and responsibility, Science, 205:277-280.
Berkowitz, Monroe, May 1979, Occupational safety and health, Ann. Am.
    Acad. of Polit. Soc. Sci., 443:41-53.
Bick, Thomas, and Kasperson, Roger E., Oct. 1978, Pitfalls of hazard
    management: the CPSC experiment, Environment, 20:30-42.
Black, S. C., and Niehaus, F., 1980, How safe is "too" safe? Int.
    Atomic Energy Agency Bull., 22:40-50.
Bogen, Kenneth T., 1980, Public policy and technological risk, IDEA:
    J. Law Tech., 21(1):37-74.
Burton, I., Kates, R., and White, G., 1978, "The Environment as
    Hazard," Allen & Unwin, London.
Cairns, John Jr., Feb. 1980, Estimating hazard, BioSci., 30:101-107.
Carter, Luther J., May 30, 1979, Dispute over cancer risk quantifi-
    cation, Science, 203:1324-1325.
Carter, Luther J., May 25, 1979, How to assess cancer risks,
    Science, 204:811-816.
Chalfant, James C., et al., 1979, Recombinant DNA: a case study
    in regulation of scientific research, Ecology Law Quarterly,
    8(55):55-129.
Chicken, John C., 1975, "Hazard Control Policy in Britain," Permagon
    Press, New York.
Conley, Bryan C., Mar. 1976, The value of human life in the demand
    for safety. Am. Econ. Rev., 66:45-55.
Covello, Vincent, and Menkes, Joshua, 1981, Issues in risk analysis
    in: "Risk in the Technological Society," C. Hohenemser and
    J. Kasperson, eds., Westview Press, Boulder, Colo.
Eichholz, Geoffrey G., Sept.-Oct. 1976, Cost-benefit assessment
    for nuclear plants, Nuclear Safety, 17:525-539.
Epstein, Samuel S., 1979, "The Politics of Cancer," Anchor Books,
    Garden City, N. Y.
Fischhoff, Baruch, 1977, Cost-benefit analysis and the art of motor-
    cycle maintenance, Policy Sci., 8:177-202.
Fischhoff, Baruch, May 1979, Informed consent in societal risk-
    benefit decisions, Tech. Forecasting Social Change, 13:347-357.
Fischhoff, Baruch, et al., Sept. 1978, Handling hazards: can
    hazard management be improved? Environment, 20(7):16-19, 32-
    37.
Fox, Renee C., Winter 1980, The evolution of medical uncertainty,
    Milbank Mem. Fund Quarterly/Health & Society, 58:1-49.
Gould, L., and Walker, C. A., eds., "Too Hot to Handle: Public
    Policy Issues in Nuclear Waste Management," Yale University
    Press, New Haven.
Greene, Mark R., May 1979, A review and evaluation of selected
    government programs to handle risk, Ann. Am. Acad. Polit. Soc.
    Sci., 443:129-144.
Hammond, E. Cuyler, and Selikoff, Irving J., eds., Oct. 26, 1979,
    Public control of environmental health hazards [45 related
    papers plus panel discussions], Ann. N.Y. Acad. Sci., 329:
    1-405.

Hapgood, Fred, Jan. 1979, Risk-benefit analysis: putting a price
     on life, Atlantic, 243:33-38.
Harriss, Robert C., and Hohenemser, Christoph, Nov. 1978, Mercury
     . . . measuring and managing the risk, Environment, 20:25-36.
Herbert, John H., et al., Jul./Aug. 1979, A risk business: energy
     production and the Inhaber Report, Environment, 20:28-33.
Inhaber, Herbert, Feb. 1979, Risk with energy from conventional
     and nonconventional sources, Science, 203:718-723.
Jacobs, Phillips, Analyzing environmental hazards, Environ. Sci.
     Tech., 13:526-529.
Jones-Lee, M. W., 1976, "The Value of Life: An Economic Analysis,"
     Martin Robertson & Co., London.
Kasper, Raphael, Aug. 1977, Cost-benefit analysis in environmental
     decision-making, George Washington Law Rev., 45:1013-1024.
Kates, Robert W., ed., 1977, "Managing Technological Hazard:
     Research Needs and Opportunities," University of Colorado
     Institute of Behavioral Sciences, Boulder, Colo.
Kletz, Trevor A., May 12, 1977, The risk equations: what risks
     should we run? New Scientist, 74:320-322.
Kunreuther, Howard, May 1979, The changing societal consequences
     of risks from natural hazards, Ann. Am. Acad. Polit. Soc. Sci.,
     443:104-116.
Kunreuther, Howard, 1978, "Disaster Insurance Protection: Public
     Policy Lessons," Wiley, New York.
Lawless, E. W., 1977, "Technology and Social Shock," Rutgers Univer-
     sity Press, New Brunswick.
Levine, Saul, Sept.-Oct. 1978, The role of risk assessment in the
     nuclear regulatory process, Nucl. Safety, 19:556-564.
Linnerooth, Joanne, Aug. 1976, Methods for evaluating mortality
     risk, Futures, 8:293-304.
Lovins, Anthony B., 1977, Cost-risk-benefit assessments in energy
     policy, George Washington Law Rev., 45:911-943.
Lowrance, William W., 1976, "Of Acceptable Risk," William Kaufman,
     Los Altos, Cal.
Martin, James G., Jan. 1979, The Delaney clause and zero risk
     tolerance, Food, Drug, Cosmetic Law J., 311:43-49.
McGinty, Lawrence, and Atherley, Gordon, 12 May 1977, Acceptability
     versus democracy, New Scientist, 74:323-325.
Mendeloff, John, 1979, "Regulating Safety: An Economic and Politi-
     cal Analysis of Occupational Safety and Health Policy,"
     MIT Press, Cambridge, Mass.
Mishan, Edward J., 1976, "Cost-Benefit Analysis," Praeger, New York.
MITRE Corporation, The Metrek Division, Sept. 1979, "Risk Assessment
     and Governmental Decision-making; Symposium/Workshop on Nuclear
     and Nonnuclear Energy Systems, Proceedings," The MITRE Corp.,
     Metrek Division, McLean, Va.
National Academy of Sciences, 1975, Committee on Principles of
     Decision-making for Regulating Chemicals in the Environment,
     "Decision-making for Regulating Chemicals in the Environment,"
     National Academic of Sciences, Washington, D.C.

National Academy of Engineering, 1972, Committee on Public Engineer-
     ing Policy. Perspectives on benefit-risk decision-making,
     "Report of a Colloquium Conducted by the Committee on Public
     Engineering Policy, National Academy of Engineering, Apr. 16-
     17, 1971," National Academy of Engineering, Washington, D.C.
National Academy of Sciences, 1977, National Research Council,
     Advisory Committee on Biological Effects of Ionizing Radiation,
     "Considerations of Health Benefit Cost Analysis for Activities
     Involving Ionizing Radiation and Alternatives," U.S. Environ-
     mental Protection Agency, Washington, D.C. (EPA 520/4-77-003).
National Council on Radiation Protection and Measurements, Mar.
     1980, "Perceptions of Risk; Proceedings of the Fifteenth Annual
     Meeting, Mar 14-15, 1979, Washington, D.C.," pp. 1-168.
Nelkin, D., ed., 1979, "Controversy: Politics of Technical Deci-
     sions," Sage, Beverly Hills.
Okrent, D., 1980, Comment on societal risk, Science, 208:372-375.
Okrent, David., ed., 1975, "Risk-Benefit Methodology and Applica-
     tions: Some Papers Presented at the Engineering Foundation
     Workshop, Sept. 22-26, 1975, Asilomar, California," National
     Technical Information Service, Springfield, VA., (NSF-RA-X-
     75-029).
Otway, Harry J., and Pahner, Phillip D., Apr. 1976, Risk assessment,
     Futures, 8:122-134.
Otway, Harry J., et al., Apr. 1978, Nuclear power: the question
     of public acceptance, Futures, 10:109-118.
Page, Talbot, 1978, A generic view of toxic chemicals and similar
     risks, Ecology Law Quarterly, 7:207-245.
Rhoads, Steven E., Spring 1978, How much should we spend to save
     a life? Public Interest, 51:74-92.
Richmond, Chester R., 1980, "The Science of Risk Assessment: A
     Summary of a Seminar Presented on Aug. 9, 1979, Oak Ridge,
     Tenn.," Oak Ridge National Laboratory, (ORNL technical
     seminar series, ORNL/PPA-80/2).
Ritch, John B., Jr., Aug. 1979, Protecting public health from toxic
     chemicals, Environ. Sci. Tech., 13:922-926.
Rowe, William D., Aug. 1977, Governmental regulation of societal
     risks, George Washington Law Rev., 45:944-968.
Rowe, William D., 1977, "An Anatomy of Risk," Wiley, New York.
Sage, A., and White, E., Aug. 1980, Methodologies for risk and
     hazard assessment: a survey and status report, IEEE Trans.
     Syst., Man, Cybern., 10(8):425-447.
Samuels, Sheldon W., Oct. 1979, Role of scientific data in health
     decisions, Environ. Health Perspectives, 32:301-307.
Schweig, Barry B., May 1979, Products liability problem, Ann. Am.
     Acad. Polit. Soc. Sci., 443:94-103.
Schwing, Richard C., and Albers, Walter A., Jr., eds., 1980, "Socie-
     tal Risk Assessment: How Safe is Safe Enough?" Proceedings of
     an international symposium held Oct. 8-9, 1979, at the General
     Motors Research Laboratories, Warren, MI., Plenum Press, N.Y.

Singer, Max, Spring, 1978, How to reduce risks rationally, Public
     Interest, 51:93-112.
Sjoberg, Lennart, 1979, Strength of belief and risk, Policy Sci.,
     11:39-57.
Slovic, Paul, 1978, The psychology of protective behavior, J. Safety
     Res., 10:58-68.
Slovic, Paul, et al., 1979, Rating the risks, Environment, 21:14-39.
Solomon, K. A., Aug. 1980, "Issues and Problems in Inferring a Level
     of Acceptable Risk," prepared for the U.S. Department of
     Energy, The Rand Corporation, Washington, D.C.
Solomon, Kenneth A., and Abraham, Stanley C., Mar. 1980, The index
     of harm: a useful measure for comparing occupational risk
     across industries, Health Phys., 38:375-391.
Starr, Chauncey, 1969, Social benefit vs. technological risk,
     Science, 165:1232-1238.
Starr, Chauncey, and Whipple, Chris, 1980, Risks of risk decisions,
     Science, 208:1114-1119.
U.S. Congress, 1980, House of Representatives, Committee on Inter-
     state and Foreign Commerce, Subcommittee on Oversight and
     Investigations and Subcommittee on Consumer Protection and
     Commerce, "Use of Cost-Benefit Analysis by Regulatory Agencies,"
     Joint hearings, 96th Congress, 1st session, U.S. Govt. Print.
     Off., Washington, D.C.
U.S. Congress, 1980, House of Representatives, Committee on Science
     and Technology, Subcommittee on Science, Research and Technol-
     ogy; Senate, Committee on Commerce, Science and Transportation,
     Subcommittee on Science, Technology and Space; and the
     American Association for the Advancement of Science, Congress/
     Science Forum, "Risk/benefit analysis in the legislative pro-
     cess," Joint hearings, 96th Congress, 1st session, U.S. Govt.
     Print. Off., Washington, D.C.
Vaupel, James W., Autumn 1976, Early death: an American tragedy,
     Law Contemp. Problems, 40:74-116.
Vaupel, James W., and Graham, John D., Winter 1980, Egg in your
     bier? Public Interest, 58:3-17.
Wildavsky, Aaron, Jan./Feb. 1979, No risk is the highest risk of
     all, Am. Scientist, 67:32-37.
Wildavsky, Aaron, Summer 1980, Richer is safer, Public Interest,
     60:23-30.
Wilson, R., Feb. 1979, Analyzing the risks of daily life, Tech.
     Rev., pp. 45-55.
Zeckhauser, Richard, and Shepard, Donald, Autumn 1976, Where now
     for saving lives, Law Contemp. Problems, 40:5-45.
Zimmerman, Burke K., Feb. 1978, Risk-benefit analysis; the cop-out
     of government regulation, Trial, 14:43-47.

APPENDIX B: FEDERAL SUPPORT FOR RISK ANALYSIS RESEARCH

There are more than 60 agencies, commissions, committees, and bureaus in the federal government concerned with health, safety, and environmental issues (see Table 1).   Although it is possible to identify at least some risk-related research in all of those offices, the purpose of this appendix is to describe briefly three major programs of government-sponsored research that address generic issues in risk analysis.   Of the three programs, that of the National Science Foundation (NSF) is the broadest in scope.   As a result, the research activities of this agency will be given special attention.

Risk Analysis Research at the National Science Foundation

The House Committee on Science and Technology, in its 1979 National Science Foundation Authorization Report (HCST, Report no. 96-61, p. 68), observed that the ability to assess and balance risks is well behind the need.   The Committee encouraged NSF, to the extent possible, to develop a program of systematic research on comparative risk analysis.   This program was assigned to the Technology Assessment Group in the NSF Division of Policy Research and Analysis, and the name of the Group was changed to the Technology Assessment and Risk Analysis (TARA) Group.

The first task undertaken by the new TARA Group was to review the literature on risk and formulate tentative conclusions about the objectives, focus, and priorities of the program.   Based on this review and advice from other parts of NSF and other institutions, an Addendum to the PRA Program Announcement for Extramural Research (NSF 78-78) was circulated to the external research community.   The Addendum encouraged researchers to submit proposals dealing with the following nine issues:

1.   How do we determine how safe is safe enough?

2.   How adequate are the data on which we depend for estimates of the risk associated with different technologies?

3.   How are implicit estimates of risk translated into decision making?

4.   How should we deal with uncertainty?

5.   What are the institutional constraints associated with decision making involving risk and uncertainty?

6.   What factors influence individual and social perceptions of risk?

Table 1. Federal Offices Concerned with Health, Safety, and Environmental Issues

I. Regulatory agencies

  A. Consumer risks

      1. Consumer Product Safety Commission
      2. Food and Drug Administration
      3. Food Safety and Quality Service

  B. Occupational risks

      1. Occupational Safety and Health Administration
      2. Mine Safety and Health Administration

  C. Transportation risks

      1. National Highway Traffic Safety Administration
      2. Federal Highway Administration
      3. Federal Railroad Administration
      4. Materials Transportation Bureau
      5. Federal Aviation Administration
      6. U.S. Coast Guard
      7. Federal Maritime Commission
      8. Urban Mass Transportation Administration

  D. Environmental and public health risks

      1. Environmental Protection Agency
      2. Public Health Service
      3. Forest Service
      4. National Oceanic and Atmospheric Administration
      5. Bureau of Land Management
      6. Fish and Wildlife Service
      7. Geological Survey
      8. National Park Service

  E. Energy risks

      1. Nuclear Regulatory Commission
      2. Economic Regulatory Administration
          (in the Department of Energy)
      3. Federal Energy Regulatory Commission
      4. Other agencies previously listed
          --Mine Safety and Health Administration
          --Materials Transportation Bureau
          --Federal Maritime Commission
          --Environmental Protection Agency

Table 1 (continued)

4. Other agencies previously listed (continued)
   --National Oceanic and Atmospheric Administration
   --Geological Survey

II. Oversight, review, and coordination agencies

A. In Congress

1. Various Congressional Committees
2. Congressional Budget Office
3. Office of Technology Assessment
4. General Accounting Office

B. In Executive Office of the President or Independent

1. Office of Management and Budget
2. Regulatory Analysis Review Group
3. Office of Science and Technology Policy
4. Regulatory Council
5. Council on Environmental Quality
6. The Administrative Conference

III. Research agencies

1. Library of Congress Research Service
2. National Science Foundation
3. National Academy of Sciences
4. National Institute of Health
5. Center for Disease Control
6. National Institute for Occupational Safety and Health
7. Agricultural Research Service
8. National Bureau of Standards
9. National Fire Prevention and Control Administration
10. Office of Naval Research
11. Alcohol, Drug Abuse, and Mental Health Administration
12. National Transportation Safety Board

IV. Other agencies

A. International relations

1. State Department
2. Department of Defense
3. National Security Council

7. How are individual perceptions of risk aggregated to
    social perceptions of risk?

8. Does society perceive that some risks are unacceptable
    no matter what the expected benefits?

9. How are equity, distributive, and other normative consid-
    erations balanced in the decision-making process?

In keeping with these stated interests, in 1979-1980 the TARA
Group made eight risk analysis awards (Table 2). Two of these
awards, to the J. H. Wiggins Company and SRI International, specifically addressed topics embedded in issues (1)-(3). At the J. H.
Wiggins Company, Lloyd Philipson and Arthur Atkisson are conducting
a comprehensive survey and critical analysis of risk assessment
methods and their employment by policy analysts and policymakers.
Using public documents, they will: (1) identify the range of federal risk management activites; (2) examine the analysis and decision processes associated with such activities; and (3) determine
the extent to which formal risk assessment methods are currently
used in risk management decision making. In the survey and analysis,
the project will focus on three major types of hazardous systems:
energy systems, public transportation and hazardous materials transportation systems, and chemical processing and storage systems.

At SRI International, Miley Merkhofer is comparing decision
analysis methods with other methods for the analysis of decision
making (such as cost-benefit analysis, cost-effectiveness analysis,
and value-impact assessment) and determining their applicability for
risk analysis. Specific objectives of the study are (1) to examine
the strengths, weaknesses, and limits of applicability of each
method, and (2) to summarize the results of the comparison in a
matrix characterizing each method along several dimensions, such
as theoretical justification, input data requirements, and ease of
use. An important element of the research plan is to develop case
studies of selected risk analysis applications.

In 1979-1980, four awards--to Clark University, The University
of California at Los Angeles (UCLA), New York University (NYU), and
the Franklin Pierce Law Center--specifically addressed topics
embedded in issues (4)-(5).

At Clark University, Robert Kates is examining generic
approaches to hazard management. The objectives of this study are:
(1) to formulate a taxonomy of technological hazards, (2) to use
this taxonomy to suggest means for improving public policy and
governmental decision making, (3) to identify generic obstacles
preventing timely and effective governmental decision making to
control or mitigate technological hazards, (4) to develop a

Table 2. NSF Risk Analysis Awards, 1979-1980

Principal Investigator Institution	Title
The J. H. Wiggins Company Lloyd Philipson	An Integrated Analysis of Risk Assessment Methodologies and Their Employment in Government Risk Management Decision-Making
SRI International Miley W. Merkhofer	A Comparative Study of Risk Analysis Approaches
Clark University Robert Kates	Methods for Improving Public Policy for Technological Hazard Management

University of California,
  Los Angeles
  David Okrent
	Alternative Risk Management Policies for State and Local Governments

New York University Rae Zimmerman	The Management of Risk
Franklin Pierce Law Center Michael Baram	An Examination of Alternatives to Government Regulation for the Management of Technological Risks
University of Maryland Douglas MacLean	Risk and Consent: Three Conceptual Models
National Academy of Sciences David Goslin	Development of Strategies for a Program of Systematic Research to Improve Risk Analysis and Decisionmaking

conceptual framework for auditing the effectiveness of hazard management; (5) to construct an initial, comprehensive theory of hazard
management, and (6) to develop methods for improved communication
of information relating to hazards and risk.

At UCLA, David Okrent is examining alternative risk management
policies for state and local governments. The objectives of this
study are: (1) to identify several risk conditions subject to local
control, (2) to quantify, wherever possible, the risk, uncertainties,
benefits, costs, and other important attributes associated with each
risk situation, (3) to provide background material on other risk
situations of concern to state and local decision makers, (4) to

examine the existing status of risk management and regulation at the state and local level, and (5) to propose and evaluate alternative risk management policies.

At New York University, Rae Zimmerman is attempting systematically to map out a management system for high-risk technologies. Through case studies and statistical analyses of crisis management and risk assessment organizations and events, the project will examine: (1) organizational and managerial linkages among the various components of the risk management system, (2) the perceptions of risk by administrators and their relationship to organizational norms, (3) the specific relationships between organizational norms and administrative structures that either favor crisis management or anticipatory risk assessment, and (4) the need for flexibility in the risk management system.

At the Franklin Pierce Law Center, Michael Baram is examining alternatives to government regulation for the management of technological risks. The objectives of the study are (1) to develop a conceptual framework for identifying alternatives to government regulation, (2) to develop baseline information on risk management strategy options, (3) to develop and apply criteria for comparatively assessing risk management strategy options, (4) to use case studies to modify the conceptual framework and criteria for assessing strategy options, and (5) to identify likely obstacles to the implementation of alternative risk management strategies.

No awards were made in 1979-1980 specifically focused on issues (6), (7), or (8): "What factors influence individual and social perceptions of risk?"; "How are individual perceptions of risk aggregated to social perceptions of risk?"; and "Does society perceive that some risks are unacceptable no matter what the expected benefits?" These questions, however, will be partially addressed as part of the Clark University project through a subcontract to Paul Slovic, Baruch Fischhoff, and Sara Lichtenstein, of Decision Research, Inc. The Decision Research team will extend their previous work on public perception of risk by examining a broader range of technologies and developing rating scales over a wider range of social groups and decision makers. The results of this empirical work will be used by the Clark University group in their managerial audit and in their attempt to develop a hazard taxonomy. In addition to the joint effort at Clark University and Decision Research, perception of risk issues will also be addressed in the UCLA and NYU projects (see above). As part of both projects, the risk perceptions and choices of hazard managers will be studied.

The final topic, (9) "How are equity, distribution, and other normative considerations balances in the decision making process?", is the focus of an award made to the University of Maryland. The

principal investigator for the project, Douglas MacLean, is conducting a critical analysis of the normative and ethical assumptions
underlying different methodologies employed in the analysis of risk.
Specific objectives are (1) to prepare an initial survey of normative, ethical, and epistemological issues in risk analysis, (2) to
examine three explicit models of consent for making decisions about
risk and safety, (3) to develop a philosophical theory of risk and
consent in centralized decisionmaking contexts, and (4) to apply
the theory in at least one important policy area. During the course
of the project, interdisciplinary working groups will be convened
to discuss conclusions.

In addition to the projects already mentioned, in 1979 the TARA
Group made an award to the National Academy of Sciences to form a
Committee on Risk and Decisionmaking. The Committee, chaired by
Howard Raiffa of Harvard University, is holding meetings and workshops concerned primarily with (1) the adequacy of the decision-
relevant data available for determining risks with a significant
science and technology component, (2) the process by which such
information is used by decision makers, (3) the institutional and
organizational factors affecting the use of risk analyses in the
decision-making process, and (4) the identification of decision-
making tools or alternative courses of action that can be used by
risk management decision makers. The objectives of the project
are (a) to assess the current state of knowledge concerning major
issues in risk analysis and the use of information about risk in
the decision-making process, and (b) to develop an agenda for
future research on the science and technology policy aspects of
risk analysis.

Risk Analysis Research at the U.S. Environmental Protection Agency

(EPA) Office of Air Quality Planning and Standards. EPA's
Office of Air Quality Planning and Standards is currently sponsoring
a major program of research on the development of quantitative risk
analysis methods. Although the program is focused on the health
risks associated with alternative ambient air quality standards, the
research is generic in nature and should be of general interest to
the risk analysis research community. The following reports are
now available:

"Reporting of Uncertainties in Risk Analysis" (H. Raiffa and
        R. Zeckhauser)
"Assessing Health Risks Associated with Ambient Air Quality
        Standards" (B. Fischhoff and C. Whipple)
"Estimation of Risk of Adverse Health Effects Associated with
        Air Quality Standards for Pollutants" (A. O. Hartley,
        K. G. Manton, and M. A. Woodbury)

"A Procedure Based on Decision Analysis for Assessing the Health
      Risks Associated with Alternative Ambient Air Quality
      Standards" (M. Merkhofer)
"A Risk Assessment Methodology for Environmental Pollutants"
      (R. L. Winkler and R. K. Sarin)
"A Conceptual Risk-Assessment Procedure" (R. de Neufville and
      M. E. Pate)
"Fuzzy Concepts in the Analysis of Public Health Risks"
      (T. B. Feagans and W. F. Biller)
"Quantitative Risk Assessment of Noncarcinogenic Ambient Air
      Quality Standards: A Discussion of Conceptual Approaches,
      Input Information, and Output Measures" (D. E. Moreau)
"Encoding Subjective Probabilities: A Psychological and Psycho-
      metric Review" (T. S. Wallsten and D. V. Budesen)
"A Proposed Framework for Quantitative Risk Assessment and
      Decision Analytic Approaches in the Review of National
      Ambient Air Quality Standards" (H. Richmond)

Risk Analysis Research at the Nuclear Regulatory Commission (NRC)

The NRC Office of Nuclear Regulatory Research has in progress
a research program on safety goal decision making. Specific activities include:

• Comparative Risk Assessment and Acceptable Risk Criteria project
  at the Oak Ridge National Laboratory to develop methods for
  addressing unacceptable and acceptable risk. A report entitled
  "Approaches to Acceptable Risk: A Critical Guide," written
  by B. Fischhoff, S. Lichtenstein, P. Slovic, R. Kenney, and
  S. Derby, is currently available.

• Research task force to formulate numerical safety goal criteria, using different technical approaches. The formation of
  the research task force and the conduct of its meetings are
  being coordinated through the Institute of Electrical and
  Electronic Engineering (IEE), with cooperation from other professional engineering societies.

• Project at Brookhaven National Laboratory to independently formulate safety goal criteria, to investigate the implications of
  such criteria, and to determine the impact of attempting to
  satisfy such criteria. Information on risk exposure and risk
  acceptance criteria from other societal activities is also
  being collected.

REFERENCES

Covello, V., and Menkes, J., 1981, Issues in risk analysis, in:
    "Risk in the Technological Society," C. Hohenemser and J. Kas-
    person, eds., Westview Press, Boulder, Colo.
Knorr, K., 1977, Policy-makers' use of social science knowledge:
    Symbolic or instrumental? in: "Using Social Research in Public
    Policy Making," C. Weiss, ed., Lexington Books, Lexington,
    Mass.
Larsen, J., March 1980, Knowledge utilization: What is it? Know-
    ledge: Creation, Diffusion, Utilization, 1(3):421-442.
Lindblom, C. E., and Cohen, D. K., 1978, "Social Research in Social
    Problem Solving: The Issues," Working paper No. 803, Institu-
    tion for Social and Policy Studies, Yale University, New
    Haven.
Lynn, Laurence E., Caplan, N., Morrison, A., and Stambaugh, R. J.,
    1975, "The Use of Social Science Knowledge in Policy Decisions
    at the National Level," Institute for Social Research, Univer-
    sity of Michigan, Ann Arbor.
Weiss, C., 1977a, Research for policy's sake: The enlightenment
    function of social research, Policy Analysis 3(4):531-46.
Weiss, C., ed., 1977b, "Using Social Research in Public Policy
    Making," Lexington Books, Lexington, Mass.
Weiss, C., 1978, Improving the linkage between social research
    and public policy, in: "The Use of Social Science Knowledge
    in Policy Decisions at the National Level, Laurence E. Lynn,
    N. Caplan, A. Morrison, and R. J. Stambaugh (1975), Institute
    for Social Research, University of Michigan, Ann Arbor.
Wildavsky, A., 1979, No risk is the highest risk of all, American
    Scientist, Vol. 67 (Jan.-Feb. 1979) pp. 32-37.

RISK/BENEFIT TRADE-OFF ANALYSIS IN WATER RESOURCES PLANNING

Warren A. Hall

School of Engineering
Colorado State University
Ft. Collins, Colorado

"For want of a nail . . ."

This little ditty contains the basic characteristics of value associated with any resource. For the most part, resources are of little value, perhaps even negative value, until they are put to beneficial use. By definition they are means to accomplish ends and are seldom, if ever, ends in themselves. To be put to beneficial use, a resource must be available when and where needed, in the appropriate amounts and in an appropriate form or "quality."

Lightning represents an enormous amount of electrical energy, yet nowhere is it utilized as a resource. Its reliability in both place and time is so poor that, large as it is, it cannot be utilized with our current technology. In fact, because it does occur at times and places where it is destructive, it would have to be considered a "negative resource" even though in its basic physics it is the same as the energy used to light this room.

Society has come a long way in the technologies it has developed for making reliable resources out of natural systems. It has done so well, in fact, that we speak of the value of nails, not in terms of kingdoms lost, but rather in cents per pound. Milk, butter, eggs, wheat, beef, bread, etc., are all "valued" as so many monetary units per quart, pound, dozen, bushel, etc. The reliability of the availability of these items has been made so high (perhaps illusory) that we need use only a single scalar to provide an index which can be used as a measure of the worth or value of the benefits obtainable from the use of these resources for the purposes intended. We then assign that index as being the worth of the resource, highly confident that this index reflects the social gains from the host of purposes for which these resources are provided.

However, this single scalar index is valid only so long as the reliability remains at its implicit high level. In reality, it required at least two scalar quantities (usually a great many scalars) to define the worth of these commodities, at least one to represent time-place reliability and the other the quantity. Lightning represents a tremendous quantity of electricity but the appropriate indices of space-time reliability puts it in the class of negative rather than positive resources.

Our logical thought processes have led to two basic proposals for dealing with this problem. One is economic, the other is statistical. The underlying idea is that under a certain set of assumptions, collectively referred to as free market conditions, potential uses and users of resources will adjust in an equilibrium type of response to the vagaries of the resource supply. Thus the effect of unreliability of the supply of a resource is simply presumed to be converted into a variable price, i.e., a single scalar index of value, which increases when the supply is low and decreases when it is high.

The statistical approach is much simpler. We have been taught (perhaps incorrectly) that, when we are faced with a vagarious phenomenon, the best measure of the results is the mathematical expectancy. This is defined by determining the probability of occurrence of each possible outcome (in this case the outcome is the amount of resource available) and multiplying it by the consequences of having that outcome. By this technique the effect of unreliability and quantity of supply are presumed to be converted into a single scalar physical quantity.

The difficulty with these approaches for water resources planning is not to be found in the basic logic of either theory but rather in (a) the degree to which the postulates required for the theory are met in the actual system and (b) the degree to which the mathematical models which we use to reflect the theory remain invarient or insensitive to changes in reliability. Both of these difficulties can create some serious problems for their blind use in water resources planning.

For example, suppose we analyze a relatively small irrigation project on this basis. We will presume that the project is so small that its production will have negligible effect on prices due to supply and demand. We will also presume that there is enough land to be irrigated, such that land will never be a constraint. Finally we will assume that water requirements as well as yields are proportional to area irrigated. All of these are commonly made assumptions.

Now let us determine the mathematical expectancy of economic returns from the project, using a stream with known statistical characteristics (mean, standard deviation, skew, etc., plus all relevant serial correlations). When we do this we reach the astounding conclusion that the maximum mathematical expectancy of the economic return will occur when no reservoir at all is provided in the project. This follows from the linear nature of the assumed production relationship between dollars worth of produce and acre feet of water.

The point that I wish to make is that the vital factor, reliability of supply, which is essential to the evaluation of every resource, is not adequately considered by either the economic approach or the statistical approach as commonly used. In the above example we combined both approaches under the reasonable hypotheses made and produced a nonsensical answer. Reliability issues were in fact totally ignored. This resulted because reliability deals not with mean values but with the characteristics of the unreliable situation which affects the result. In all cases these include the deviations from the mean. Furthermore, in most resource situations it is even more concerned with the specific sequences of deviations from the mean which can be reasonably expected to be encountered in the future.

The most obviously ignored sequence in our simple example was the sequence of flows during the crop production season, and how well these match the time of need by the growing plants. In our linear model, the production from an entire year's streamflow will be about that which results from the minimum ratio of supply to crop needs encountered as the season progresses. In the actual system the water in excess of this amount produces virtually nothing, while the model used presumes every unit of water produces the same amount of product regardless of when it is available.

Rather less obvious are the reliability requirements over a number of seasons. It is fairly clear for perennial crops, such as fruit trees, which require a number of years of growth before economical yields are produced. Suppose water is available to such a grower 3 years out of 4. Obviously he will be very fortunate to avoid a perpetual state of bankruptcy. It does not suffice to express the "mathematical expectancy" of his annual return as 75% of the average return for a long sequence of perfect years. In fact the probability of 0.75 has no significance whatever for this particular problem.

We are not concerned with the probabilities of a given daily flow, monthly flow, annual flow, or even a twenty- or fifty-year average flow. The concern is with the probability of sequences of flows, rather indefinite in duration, which could result in the loss of some or all of the rest of the resources (investment largely) needed to produce the perennial crops. Of critical importance are the probabilities associated with sequences of streamflows of one, two, three, etc., periods of flow, up to twenty to thirty years or more duration.

Even with annual crops, the concern will be for sequences of flows substantially longer than one growing season. Although the consequences of shortage are not as catastrophic as those suggested for perennial crops, there still will be major investments which must be made, and which will not be made by prudent investors unless the reliability is sufficiently high for durations corresponding to the order of two investment recovery periods. As a minimum, any statistical or economic analysis which does not recognize this limitation may be seriously in error.

Standard water resources planning procedures recognized these dual criteria for value (reliability-benefit) in an implicit way. The water or energy yield of a set of project facilities was defined to be that annual quantity of water or energy which could be delivered when required, without fail if the historical stream-flow record were to repeat itself exactly, subsequent to reaching a use level corresponding to that amount. Since in most cases the historical streamflow record was of the order of 50 years duration, this provided a reasonably high reliability for most practical purposes.

The concept of sequences of events is implicit in this definition of resource availability. The presumption is that the sequences of events that actually did occur represented the best available samples of what might be reasonably expected to occur in the future. In fact, they were the only available samples.

There has been a number of serious misconceptions resulting from the use of these "Rippl diagram" or "mass balance analysis" procedures. The least serious of these is the miconception that the engineers making the analysis were assuming that "history would repeat itself exactly." This is of course absurd, but it has frequently been used as an argument against continued use of the procedures. The only assumption that was made is that the historical sequences represented the best available samples of the various sequences of flows which could happen in the future.

A second misconception is much more serious, in that it has encouraged us to use a single scalar quantity representing the level of resource availability, omitting the all-important implicit levels of risk associated with that level. When we state that a project will provide so many acre-feet of water per year, so many kilowatt-hrs per year of electricity, reduce concentrations of contaminants to such and such a level, we seriously mislead the uninitiated

(including some lawmakers and an overwhelming majority of the general public) by omitting any references to the critical concern for
reliability. A million acre-feet of water per year is valueless if
the reliability of its availability when and where needed is too
low to permit beneficial use. The same is true of electricity, flood
protection, water quality control, recreation, and almost all other
beneficial purposes associated with water resources development.

So long as the procedures and practices remain standard, using
the long-term mass balance analysis, the miconception does little
harm. The trouble comes with the suggestions and arguments for
changes in the procedures.

These suggestions and arguments do have validity. We now have
computer technologies which allow for much better mathematical modeling and for including far more factors of interest and concern in the
analysis. Unfortunately many of the proposals for change either do
not recognize the critical importance of reliability of risk avoidance in the basic concept of a resource or they tacitly (but erroneously) assume that reliability is properly considered by using
mathematical expectation and/or economics of supply-demand, in an
oversimplified manner.

To correctly reflect risk using mathematical expectation, for
every combination of sequences of events which would influence outcome (utility of the resource for purposes intended) we must assess
the corresponding probability and combine the total with due respect
for the laws of probability of independent and non-independent
events. This is a virtual impossibility in most cases where reliability involving various combinations of streamflows in sequence
is concerned, except by Monte Carlo type simulation as is done by
standard practice.

To correctly represent supply-demand effects in terms of economics, the levels of reliability as well as levels of quantity
involved must be simulated. We are not concerned here with an
instantaneous market concept as a quantity of goods available for
sale being in equilibrium with a quantity of goods desired to be
purchased. As we have seen earlier, this value is essentially zero
for an unreliable supply over time, and remains very low until the
reliability reaches relatively high levels. I know of no instance
where the economic price elasticity of this vectoral nature of
resource supply has been determined, and I suspect it would be quite
impossible to accomplish with the present state of economic science
and data availability.

This is not an argument for status quo in project analysis.
It is an argument for caution, however, in the development and

evaluation of the validity of any alternate procedures that might
be suggested.

It is clear that a trade-off between nominal resource level
and the reliability of that level is a common economic-social practice. There is no risk-free resource level. This suggests that
there is always a trade-off involved between nominal resource level
and reliability levels (plural). This trade-off can and should be
optimally determined. The results of historically recorded mass-
balance analysis are not necessarily optimum or even close thereto.
It has been an expedient that has served us well, and which should
not be abandoned without careful consideration of the validity and
utility of alternative procedures. However, we cannot argue that a
procedure which only implicitly considers hydrologic risk, and does
so without even evaluating that risk, will automatically produce
the optimum trade-off between resource quantity and reliability
levels.

The problem is not at all simple. There are two basic concerns
which must be treated carefully. The first and easiest to handle is
the probabilities of sequences of streamflow quantities. The
emphasis must be on sequences of streamflows of various durations
up to and including all durations which might affect the purposes
of the resource. If thirty years of uninterrupted service is important to the particular intended use of the resource, then any equally
likely sequence of thirty years duration is one event for that use.
It is not thirty annual events nor 360 monthly events but one single
event. To evaluate the associated risk, one would need a fairly
large number of such "events," say 50, or the equivalent of 1500
years of streamflow record.

Under some rather restrictive hypotheses, which must be reviewed
for each case, it is possible to generate "synthetic hydrologies"
whose statistical characteristics are virtually indistinguishable
from those of the actual historical record (best available sample).
When the hypotheses are valid, it is then at least possible to
estimate some parameters related to the reliability of supply provided from a reservoir for any desired duration of sequences.

The problem of quantification is made a bit difficult in that
the reliability itself is a multi-component "vector" or quantity
with an infinite number of components. First of all there is the
reliability measure associated with each and every sequence duration which, if significantly altered, would also significantly alter
the consequences. For example, the duration, magnitude, and temporal
distribution of a resource shortage all are significant elements
which cannot be conveniently combined into a single scalar number
whose probability of non-exceedence might then be calculated. Furthermore, even if this could be done, the non-exceedence probability

of a shortage of, say 10% does not have the same importance with respect to ultimate consequences as does the nonexceedence probability of a 20% shortage. Each is a risk to be avoided, but each would have a different "trade-off" at optimality with the nominal beneficial levels provided for use. To make matters more difficult, these and the infinite spectrum of other percentages of shortage are not at all independent but rather are interdependent in varying degrees depending on the strategies being utilized to operate the system in the face of a potential storage. The same difficulties are encounttered but to a greater extent with all the other dimensions of risk avoidance such as sequence, duration, etc.

We can utilize the Monte Carlo "equally likely" hydrological sequence-generation techniques, if the required assumptions are met satisfactorily, as a means of evaluating most of these risk characteristics for a given physical resource development system, for its objectives and purposes, and for a given fixed operational policy. These could then be reevaluated for any proposed change in the given physical system or the given operational policy to determine the marginal response in all related risk parameters for the given marginal change in system or policy.

Obviously considerable judgment would be needed in the selection of the initial trials if an impossible number of evaluations is to be avoided. For example, it might be most expedient to begin with an analytical procedure, such as the classical mass balance analysis, which is reasonably sure to provide as much or more security against hydrologic risk of all types as would be needed, and then test designs and operational policies which reduce costs but must also minimize resulting adverse effects on reliability.

We could continue with this approach but there are some additional considerations which must be identified and treated. It was previously pointed out that the probabilities of various levels of deficiency and their distribution and duration are not independent but in fact are at least partially dependent, depending on the operational policies primarily, and to a lesser extent on design of facilities. Under most circumstances this can be converted to an analytical asset since, to the degree that they do correlate highly, this will allow selection of one factor and its risk parameters as the single index of all factors for purposes of the trade-off analysis and search for the optimum.

However, the technique must be used with caution. Operational procedures can sometimes be designed which will simply shift the risk from one category to another. This may in fact be desirable, but it violates the assumptions of interdependent implicit in the use of a single risk index.

This leads us to the most important aspect of risk-benefit analysis, that of assuring that the models of benefit production are reasonably well tailored to the models being used to evaluate risk. As indicated previously these are not at all independent. The uses to which a resource can be put are economically and socially determined by the levels of the risk parameters associated with the resource availability in time and place. Industrial use of water may appear to be a high-economic-value use, but most industrial uses will not be undertaken unless the supply reliability is made relatively high. The same is true, to a greater or lesser extent, for virtually all other uses. Mathematical models of the consequences of resource use are not valid unless minimum reliability requirements for undertaking that use are met, and unless any effects of reliability above that minimum on productivity are correctly modeled.

A practical example of this problem was reported in a symposium sponsored some years ago by the Rockefeller Foundation on the use of high-yielding varieties of crops. I regret that I have lost the exact reference. Basically a Central American nation was importing corn even though most of the farmers were still using the old native, low-yield varieties. Analyses in the classical sense showed that the farmers could increase their economic income about ten-fold if they switched to high-yield varieties. It was also found that the farmers knew about these miracle seeds, but refused to use them. The refusal was attributed to ignorance and tradition since the economic benefits were obvious. However, risk (though low) was not considered. When I calculated the economic consequences of a single crop failure using the new seeds, I discovered that the farmer would be in ever-increasing debt for the rest of his life thereafter.

On a national average basis, the risk of failure on a grand scale was sufficiently low that the new seeds were well worth the risk. However, from the individual farmer's point of view, ignorant and non-quantitative as it may have been, his personal analysis was clearly the correct decision. Crop failure with native varieties was virtually unheard of. The new seeds were clearly more vulnerable to disease, insects, and drought.

If the user won't use the resources, or won't use them as provided for in the mathematical models used in the analysis, then the analysis is wrong, however sophisticated those models might be.

This brings up yet another issue in risk-benefit analysis. The security of supply must not only exist in design and operational policy, but there must also be credibility of that security in the mind of the user. In the above example, it would appear that the consequences of risks of failure could be easily and completely eliminated by the simple expedient of a governmentally or privately sponsored insurance program. This is not necessarily so. Many of

the farmers involved have an inherent and perhaps well-founded distrust of such governmental or private promises (for that is all they
would be). If so, then for all practical purposes, the necessary
credibility of the reliability promised would be lacking and the
farmers involved would still not change their practice.

Closer to home, there are many persons who strongly advocate
elimination of the water rights system (which is the basis for reliability of supply for most water uses involving diversion), substituting therefore a system of periodic transfers to "highest and
best use" as determined by a set of economic supply-demand transac-
tions. It is obvious that this reduces the reliability of supply
for any use or user. The reduction in reliability would be virtually
complete for a daily auction system.

Thus we see we must consider what may appear to some to be
redundancy. That is, we must not only consider the reliability of
supply for various uses but must consider the assurance of that
reliability level as well.

In the case of the Central American farmers cited earlier, even
with a governmentally sponsored insurance system, there was still
lacking the assurance that the level of reliability provided by insurance would in fact be met. Governments in that area are not
characteristically stable, nor are administrative policies reliable
even within a stable government.

It is basically this instability of policy which led to the
creation of water rights as essentially property rights based on
the protection of the Constitution rather than on administrative
authority. The latter might achieve some people's definition of
"highest and best use" but it is doubtful that it would have provided the credibility by the administrator's assurances of the level
of reliability promised, particularly if the administrator was a
political appointee of the opposite party to the users.

Once again, this is not to argue that the existing water rights
system is perfect and should not be changed. What I do argue is
that any changes must give due consideration to the absolute necessity of both a known level of reliability of supply and a level of
assurance that that level of reliability will in fact materialize.

A known level of reliability is the key to the beneficial use
of any resource. Assurance of that level of reliability is the key
to the materialization of the expected use and corresponding social
benefits of the resource. When analyzing the development of new
water resources for whatever purposes, or when analyzing modifica-
tions of policies for existing uses and users, both of these aspects
of risk-benefit analysis must be properly identified and provided

for in the mathematical and descriptive models used in the analyses.

> For want of a nail the shoe is lost
> For want of the shoe the horse is lost
> For want of the horse the rider is lost
> For want of the rider the battle is lost
> For want of the battle the kingdom is lost

The value of a resource is always to be found in the degree of the reliability of its availability when and where needed, and in the quantities and qualities needed, and over substantial periods of time. Furthermore, uses of resources requiring any particular level of reliability will not in fact be undertaken unless the user has a high degree of <u>assurance</u> that the actual reliability in the future will not fall below the necessary level for that use.

MINIMIZING RISK OF FLOOD LOSS IN THE NATIONAL FLOOD

INSURANCE PROGRAM*

Malcolm Simmons

Environmental and Natural Resources Policy Division
Congressional Research Service, The Library of Congress
Washington, D.C.

INTRODUCTION

This paper will discuss the concept of risk in the National
Flood Insurance Program enacted in 1968. The present program is
different from the 1968 program in one important way--it contains
incentives that make program participation nearly mandatory. The
mandatory nature of the program was a conscious national policy
decision to limit flood-plain development and thereby reduce the
taxpayers' burden of flood relief payments. The alternative was
the previous situation in which flood relief payments had continued
to escalate because of uncontrolled flood-plain development.

The National Flood Insurance Program is based on the economic
principle that occupants should pay the full economic costs arising
from their use of flood hazard areas. Compulsory flood insurance
with actuarially determined premiums places the financial burden of
flood loss on property owners rather than on taxpayers, thereby
encouraging more efficient location decisions which avoid the risk
of flood loss in the future.

The following will review the theory of individual decision and
the implications for mandatory flood-plain insurance, the economic
theory of mandatory insurance, the history of the National Flood
Insurance Program, the nature of the federal involvement, the

---

*The views expressed in this paper are those of the author and do
not necessarily reflect the views or positions of the Congressional
Research Service.

requirements for entrance into the program, risk coverage in the
program, and the mapping studies in the program.

INDIVIDUAL DECISION

A basic question in the development of the National Flood Insurance Program was how to get individuals to participate. There are
essentially two approaches: the carrot approach, where incentives
attract voluntary participation; and the stick approach, where negative sanctions for nonparticipation would induce participation. The
original program authorized by the 1968 National Flood Insurance Act
used the voluntary approach, only to discover that the incentives
did not attract sufficient participation in the program. In 1973
the program was amended to allow the use of a mandatory approach.

Individual decision is important from the policy standpoint
because the consequences of this behavior must be accepted by the
nation as a whole. If, for example, economic losses result from
this behavior, then the taxpayer at large must pay for this loss
through relief payments.

There are three major factors that influence individual decisions to participate in insurance programs to avoid risk: perception, affluence, and personality. In regard to perception of the
flood risk, inadequate or unavailable information on flood flows
and frequencies may affect behavior. Another possibility is that
individuals may have information of flood flows and frequencies,
but not be able to utilize such information. One example is an
individual who knows the probability of a flood event, but has never
experienced flooding. This individual would be less likely to purchase flood insurance than one who experiences a flood event every
five or ten years. Kunreuter [1978, p. xi] characterized this
phenomenon in the context of the National Flood Insurance Program,
where few people have had the exposure to the low-probability event
of the 100-year flood. He states that individuals will not protect
themselves voluntarily against such a low-probability event.

A strong correlation exists between affluence and the desire
to reduce risk. Wildavsky [1979] elucidates this theory in the
societal context in his paper Richer is safer. Kunreuter [1978,
p. 242] found that, as an individual's income increased, the chances
of purchasing a flood insurance policy also increased somewhat. He
also found a positive correlation between educational level and age
and the desire to avoid risk through purchase of flood insurance.

Personality also may affect risk-taking behavior. A risk-
avoider would be more willing to participate in a flood insurance
program than a risk-taker.

When the flood insurance program was voluntary, very few individuals purchased flood-plain insurance. Even though coverage was highly subsidized by the federal government, less than 3,000 out of the 21,000 flood-prone communities in the United States entered the program during its first four years of operation, and fewer than 275,000 homeowners voluntarily bought a policy [Kunreuter, 1978, p. 6].

Kunreuter [1978, p. xi] concludes that relatively few people will expend effort and money to avail themselves of needed protection--either through the development of loss-prevention techniques such as sound land use and control measures in flood-prone areas or through the purchase of flood insurance.

This individual behavior in the days of voluntary participation in the National Flood Insurance Program was particularly disturbing because the federal government provided high subsidies for premiums on existing homes. This indivudal behavior had social importance because, if people would not protect themselves voluntarily against the probability of a flood event, then society would continue to bear a large proportion of the costs following the disaster. When individual participation in the program remained low and federal expenditures on flood relief continued to increase, the Congress sought to reverse both these trends by requiring mandatory participation in the program.

ECONOMIC THEORY OF MANDATORY INSURANCE

Two important theorists for mandatory flood insurance are Gilbert White and John V. Krutilla. As early as 1945, White noted that building dams and dikes without restraining further occupancy of the flood plain would invite greater losses upon the occurrence of a storm exceeding the design limits. White and others have argued that structural projects, such as dams and dikes, should be accompanied by nonstructural measures to restrict new development in flood plains. Flood-plain insurance is an example of such a nonstructural measure.

Krutilla advanced a strong argument for mandatory flood insurance in 1966. He pointed out that the flood relief program existing at that time produced an allocation of the benefits and costs of flood-plain occupancy that encouraged uneconomic use of the flood plain. The taxpayers paid the costs of the construction projects to protect the flood plains, yet the dwellers in the flood plains were the beneficiaries. Where the taxpayers bear these costs, over-investment in flood control will probably occur, which in turn will encourage excessive occupancy of the flood plain.

The theory behind the present program is that the beneficiaries of flood control benefits should pay the costs. A mandatory flood insurance program would more effectively relate these benefits to costs and thereby promote a more efficient use of flood-prone areas. A requirement that each new flood-plain occupant pay a flood insurance premium that covers all flood-related costs forces the occupant to compare the benefits of locating in a flood hazard area with the full costs of this location. Actuarial flood insurance premiums serve as the mechanism for efficiently pricing location decisions in the flood plain.

The mandatory flood insurance program, however, does not affect past decisions to locate in the flood plain. Those already in the flood plain will remain until structures are destroyed, depreciated, or become obsolete.

Beyond this theory of promoting the efficiency of flood-plain land use are the problems of actual development of the program. The basic problem in any democratic society is the acceptance of a mandatory program. Beyond this problem lie other problems such as the administration of the program, the relationship of the public sector and the private sector, and the conversion from a voluntary program.

HISTORY OF THE PROGRAM

Many changes have occurred in the National Flood Insurance Program, but the most important was the change from the voluntary program of 1968 to the mandatory program of 1973. This section outlines some of the changes in the program since its inception.

Under the original or regular program as enacted by the 1968 National Flood Insurance Act (Title XIII of the Housing and Urban Development, P.L. 90-448), insurance was not available until a detailed and time-consuming flood insurance study was completed so that insurance companies could establish actuarially sound rates and could determine and map the level at which new construction would be reasonably safe from flooding. Because this requirement severely restricted the entrance of communities into the program, the Congress amended the 1968 Act in 1969 (P.L. 91-152) to provide an emergency program. Under the emergency program the federal government subsidized the sale of flood insurance in a community as soon as its application was accepted and before the required technical studies were conducted. Originally established for a two-year period ending at the end of 1971, subsequent amendments extended the emergency program.

Communities in the emergency phase were eligible for insurance, but only half the program's total limits were available. Coverage in the emergency phase was available at highly subsidized premium

rates. After completing the necessary requirements, communities
could enter the regular program under which the full limits of coverage were available at acturial rates.

Until 1973, participation in the National Flood Insurance Program was voluntary. Relatively few property owners, however, had
voluntarily entered the program. Instead, unwise development of
flood plains had continued, and the federal government was paying
higher and higher annual cost for flood disaster relief. Since individuals were reluctant to enter the National Flood Insurance Program,
the federal government enacted the Flood Disaster Protection Act of
1973 (P.L. 93-234), which for all practical purposes made participation in the program mandatory.

The 1973 Act provided a number of strong incentives for the purchase of flood insurance and for community participation. Communities not participating in the program would forego eligibility for
purchase of flood-plain insurance at federally subsidized rates,
disaster relief in the event of a flood, and federally assisted or
guaranteed loans for new construction or mortgages on existing
buildings.

The 1977 amendments to the flood insurance program contained in
Title VII of the 1977 Housing and Community Development Act responded
to criticisms of the mandatory nature of the program and insufficient
coverage in the emergency phase. The 1977 amendments removed requirements that prohibited property owners from nonparticipating communities in designated flood-prone areas from receiving loans from federally insured or regulated private lending institutions. Also,
communities in the emergency phase could secure a basic level of
flood insurance coverage even though the necessary flood hazard elevation and actuarial rate studies had not been completed, but as long
as the community had adopted minimum flood-plain management requirements. For communities in the regular program, the 1977 amendments
permitted substantially higher "second layer" coverage limits than
under prior law.

NATURE OF THE FEDERAL INVOLVEMENT

The federal government is presently the sole administrative and
financial agent for the National Flood Insurance Program, and in this
capacity it assumes the financial responsibility for the flood losses
and premium payments. The Flood Insurance Administration, originally
under the Department of Housing and Urban Development (HUD) and now
under the Federal Emergency Management Agency (FEMA), has contracted
with the EDS Corporation for centralizing the policy servicing and
claims review process.

Before 1978, however, the FIA managed the National Flood Insurance Program in a cooperative effort with a pool of over 120 private insurance companies which comprised the National Flood Insurers Association (NFIA). Companies that participated in the association as risk-bearers committed risk capital and shared in the profits and losses of the pool's operation. Other companies participated on a non-risk basis, acting as fiscal agents for the pool.

The federal government fulfilled its responsibilities in this cooperative effort by subsidizing the private industry pool through premium equalization payments and by providing excess loss reinsurance to assist the industry in the event of catastrophic losses. The industry purchased the excess loss coverage by paying reinsurance premiums into the fund.

The role of the federal government in the National Flood Insurance Program has been a continuing source of controversy. In the original federal/private-sector partnership developed under the 1968 Act, the private insurance company pool generally sought a greater role in the administration of the program. The 1968 authorizing legislation, however, permitted the federal government the option of assuming a more direct administrative and financial responsibility than was the case in the original partnership between the FIA and the National Flood Insurers Association (NFIA).

The partnership between the FIA and NFIA dissolved at the end of 1977 when the federal government availed itself of the provisions in the 1968 authorizing legislation to assume more responsibility. An impasse had developed between the two organizations because of NFIA's desire to expend funds without prior approval of the federal government. Objecting to this posture, the Secretary of the Department of Housing and Urban Development (HUD)--in which FIA was then located--assumed more administrative control of the program. Although NFIA sought a temporary restraining order in this HUD action, the United States District Court ruled that the Secretary's authority was within the authority of the Act. After NFIA and HUD were unable to reach an accord, HUD chose the EDS Corporation to replace NFIA. The major differences between NFIA and EDS is that EDS acts purely as a financial _agent_ of HUD, assuming no financial responsibility for flood losses under the program, and EDS has centralized the policy servicing and claims review process.

REQUIREMENTS FOR THE REGULAR PROGRAM

In order to qualify for the regular program of the National Flood Insurance Program, a participating community must adopt certain land use and other regulations in the emergency program to reduce flood-plain loss. Adoption of these regulations is dependent upon community mapping studies prepared by the FIA.

Initially, in the emergency phase, the community must certify that it will regulate certain aspects of future construction. The minimum standards for meeting this certification requirement are:

- A building permit system that includes reviewing permits to assure that any known flood hazard is considered.

- Requirements for anchoring and flood-proofing structures which will be built in the known flood-prone area.

- Review of subdivision proposals to assure that they will minimize flood damage.

- Requirements that new water and sewage systems and utility lines will be safe from flooding.

During the emergency program, the FIA prepares two mapping studies for participating communities: the Flood Hazard Boundary Map (FHBM) and the Flood Insurance Rate Map (FIRM). The 100-year flood level is the determinant of the flood-plain area having special flood hazard areas superimposed on community base maps. Their official purpose is to designate which properties are subject to the mandatory requirement to purchase flood insurance. The FIRM is a follow-on mapping study to the FHBM which depicts the elevation and width of the 100-year flood-plain and the differential zones of risk within it.

Within six months after receiving the FHBM, a community in the emergency program must adopt the following controls over the hazard area:

- Require for new and substantially improved residential structures the elevation of the lowest floor (including the basement) to at least the 100-year flood level.

- Require for new or substantially improved nonresidential structures elevation or complete flood-proofing for protection against at least the 100-year flood level.

- Prohibit in the floodway new development that would cumulatively raise the flood water level more than one foot, expansion of existing structures, and fill or encroachments unless offset by stream improvements compensating for reductions in the carrying capacity.

- Require for all new construction in coastal high hazards areas that structures are elevated above the floodtide level on anchored piles.

With the FIRM complete and the flood-plain land use and other control measures enacted, a community may enter the regular program.

RISK COVERAGE

The ultimate success of the National Flood Insurance Program depends upon the development for new construction of full acturial rates reflecting the real risk of loss in a particular flood-plain location. The 1968 authorizing legislation for the program permitted not only the use of acturial rates but also subsidized rates. The subsidized rates were viewed as an inducement to prospective pur-chasers of insurance. The difference between actuarial rates and subsidized rates represent the federal subsidy.

In 1977, this subsidy was two-thirds of the program, with pre-miums paid by policyholders covering only about one-third of the total costs of the program [Powers and Shows, 1979]. The two prin-cipal costs of the National Flood Insurance Program are payment of covered losses and the costs of the flood hazard studies. In 1977, total costs of the program were $612 million, of which $211 million were for payment of flood losses and $247 million for federal studies of flood-prone areas. The remaining $154 million were for expenses for selling and maintaining policies, insurance reserves, federal administrative expenses, and miscellaneous.

The amendments to the flood control program in 1969 and 1973 induced more communities to participate by increasing risk coverage. The 1969 amendment initiated the emergency program and permitted a maximum amount (first layer) of flood protection, regardless of risk, at federally subsidized rates. The 1973 amendments provided strong incentives for community and individual participation in the program. In August 1978, 16,000 of the 20,000 estimated flood-prone communities were signed up for the program. Of the participating communities, about 13,300 were in the emergency program [GAO, 1979].

The 1973 amendments also raised the limits of coverage. The maximum for the first and second layers of coverage for single-family residence under the Emergency Program was raised to $35,000 each, thus providing an overall maximum amount of $70,000 protection. Before the 1973 amendments, first-layer coverage was limited to $17,500, with an additional layer of $17,500 available at actuarial rates once these rates were established for a community.

The 1977 amendments to the program increased second-layer (regular program) coverage for residential and other types of struc-tures. In family residences, for example, second-layer coverage was increased from $35,000 to $150,000, thus providing a total maximum amount of $185,000 protection. Table 1 presents the available limits

Table 1. Available Limits of Coverage for Existing and
New Construction Under the Regulation Program

	First Layer (Emergency)	Second Layer (Regular)	Total	Maximum Req'd under the Act
Single Family Residential	35,000	150,000	185,000	70,000
Other Residential	100,000	150,000	250,000	200,000
Small Business	100,000	150,000	250,000	200,000
Churches and Other Properties	100,000	150,000	200,000	200,000
Contents				
Residential	10,000	50,000	60,000	20,000
Small Business	100,000	200,000	300,000	200,000
Churches, Other Properites per unit	100,000	100,000	200,000	200,000

Source: Powers and Shows [1979].

of coverage for existing and new construction under the regular program.

MAPPING PROBLEMS

The 1968 authorizing legislation for the program required FIA
to produce two sets of maps: the Flood Hazard Boundary Maps (FHBM)
and the Flood Insurance Rate Maps (FIRM). The information produced
in the first set is required in production of the second set of rate
maps.

The slow pace of completing rate maps required for entrance
into the regular program of the National Flood Insurance Program has
become a real obstacle to the success of the program. Because of
the bottleneck situation that the requirement for rate map completion

could produce, the 1977 amendments attempted to alleviate this problem. The amendments provided that communities in the emergency phase
could receive a basic level of flood insurance coverage even though
the FHBM's and FIRM's had not been completed.

The 1968 authorizing legislation for the program requires that
all communities participating in the National Flood Insurance Program
must be in the regular program by 1983. The time required for the
mapping procedure--generally four years--and some technical problems
in the process indicate that this deadline will not be met by all
participating communities. By October, 1978, fewer than 3,000 of
the 16,000 participating communities were in the regular program.
FIA has stated that, although all the rate maps would not be completed by 1983, all of them would be started by then [GAO, 1979].

Initially, must of the mapping was conducted by federal agencies.
The FIA, however, has shifted much of the study and mapping work to
private firms. According to FIA, the federal agencies did not have
the resources to handle the workload of the accelerated mapping program, and FIA could control better the work schedules of private
contractors [GAO, 1979].

Many communities wanting to move ahead in flood-plain regulation are confounded by inadequacies in the FHBMs in developing land
use controls. Usually the crude approximations of flood hazard
contained in FHBMs are unsuitable for land use regulation--a requirement for entry into the regular program. Within six months of receiving the FHBM, a community must adopt the necessary land use controls for new construction. The inadequacy of the FHBMs often makes
this task impossible. Although most communities have received their
FHBMs, many were dissatisfied with the results. As of November 1,
1975, appeals were filed on 2,884 FHBMs. Of these appeals, 1,249
resulted in revisions of the maps [Platt, 1976].

In contrast to the FHBMs, the FIRMs usually contain adequate
information from which to develop land use regulations. Communities
may use the FIRMs to develop land use regulations, but the long lag
time between completion of the FHBM and completion of the FIRM makes
this an unattractive option to a community desiring to enter the
regular program as quickly as possible. Many communities are in a
quandary as to whether they should develop a land use regulation
program based on an inadequate FHBM, or delay entrance into the
regular program by waiting to develop land use regulation on the
yet-to-be completed FIRM.

Many communities are also dissatisfied with the FIRMs that have
been completed. The General Accounting Office [GAO, 1979] has identified four areas of criticism:

- Maps were inaccurate and out-of-date.

- Engineering firms doing the mappings were unfamiliar with local conditions and did not seek technical input from local officials.

- Map scales were too large to be of use.

- Map scales did not show topographic information for land areas in the flood plain.

The technical problems and delays in completing the FIRMs for many communities have resulted in an important insurance rate coverage problem in the National Flood Insurance Program. The problem is that actuarial rates cannot be charged until they are calculable, but calculation is contingent upon completion of the FIRM. A 1973 amendment attempted to alleviate this situation, but in turn created another potential problem. The amendment changed the effectiveness date for actuarial rates for new construction to December 31, 1974, or the effective date of the initial FIRM, whichever is later. The FIA has interpreted this amendment to mean that subsidized rates are available to all new construction started before completion of the FIRM. This situation contrasts with the regulations before the 1973 amendments, when new construction started before completion of the FIRM was ineligible for any flood insurance. The FIA interpretation of this amendment could create, according to one expert, a "boom" in new flood-plain construction as developers try to take advantage of the subsidized rates before the FIRM is completed [Platt, 1976]. Only if participating communities are cautious in the granting of building permits can this undue flood-plain development be controlled.

A final problem with the mapping studies in the changing nature of the flood plain. While an FHBM and a FIRM may provide the necessary information for the development of a flood-plain insurance program at one point in time, in the future this situation may change. For example, development of a river section upstream of the flood plain will usually produce higher-volume flows downstream in the flood plain. These larger average flows will change the boundaries of the 100-year flood and produce the need for modification of the FHBM and FIRM.

The FIA is now conducting a study to be completed in 1980, on ways to overcome some of the problems associated with the FHBMs and the FIRMs.

CONCLUSION

From a national policy standpoint, the theory of mandating flood-plain insurance in order to decrease costs to the taxpayers

at large appears to be a valid one. The combination of land use controls in the flood plain and use of full actuarial rates are the ultimate goals in the program which, if accomplished, should validate its
existence through reduced economic losses caused by flood damage.

The benefits from the program in this form, however, are in the
future. As an interim measure, federal subsidy of insurance premiums
for existing and new construction has performed the important function
of inducing more participation in the program. Important problems
with the mapping studies have slowed the rate at which the participating communities enter the stage of the program where land use controls
and full actuarial rates are in effect. Working out these problems
in the program would enable it to live up to its potential.

REFERENCES

GAO, U.S. General Accounting Office, 22 March 1979, letter from Henry
    Eschwege, Director, to Patricia Harris, HUD Secretary.
Kunreuter, Howard, 1978, "Disaster Insurance Protection: Public
    Policy Lessons," Wiley, New York.
Platt, Rutherford H., July 1976, The national flood insurance pro-
    gram: some midstream perspectives, J. Am. Inst. Planners.
Powers, F. B., and Shows, E. W., June 1979, A status report on the
    national flood insurance program--mid 1978, J. Risk & Insurance.
Wildavsky, Aaron, 1979, Richer is safe, paper included in testimony
    before U.S. Congress, House Committee on Science and Technology,
    Senate Committee on Commerce, Science, and Transportation, Joint
    Hearings on "Risk/Benefit Analysis in the Legislative Process."

WHAT KIND OF WATER WILL OUR CHILDREN DRINK?

David Okrent

Chemical, Nuclear and Thermal Engineering Department
University of California
Los Angeles, California

This evening I shall try to look briefly at how risk-benefit analysis might be applied to some of the things that can affect the quality of our drinking water as it relates to health. This will by no means be a complete review. And my comments may well reflect the fact that I am a novice with regard to questions of water quality and waste disposal, and am not familiar with all that the Environmental Protection Agency (EPA) and others have published which relates to the matter. So, I will be giving some early impressions.

First, what is the magnitude of the problem? In fact, my feeling is that we really don't know, and that priority should be given to obtaining a better handle on the matter. For example, one might ask:

What are the average effects on health of the currently used drinking water? Where do health effects occur which are much larger than average, how large are they, why do they occur? What are the uncertainties and what are the gaps in our knowledge in this regard? What constitutes adequate knowledge and when and how can we get it?

What are the potential future effects on drinking water, and thus, on health, of the wastes which are dumped, or otherwise disposed of, in the past? Has this been quantified? Can it be? In assessing health effects arising from these wastes, do we need to consider other uses of water, e.g. for irrigation?

How should EPA, the states, and other interested parties judge whether the controls that EPA has promulgated or is planning to

promulgate on the disposal of hazardous wastes and on drinking
water quality are appropriate, correct, necessary and sufficient,
. . . you choose the adjective.

How can we estimate the potential future effects on health due
to contamination of drinking water not only due to direct pathways
from new coal mines, new synfuels operations, etc., but also from
synergistic effects? [EPA, 1980a]. Are we likely to develop a need
to use contaminated water resources for human consumption in some
parts of the country because of an essentially total commitment of
available water resources in the region?

Let me go back to the first question pertaining to the magnitude of the problem. Epidemiological studies and animal studies
both have limitations. However, about a dozen epidemiological
studies almost all show an association between cancer rates and
some aspect of drinking water, particularly organic contaminants
[Kimm, Kuzmack, and Schnare, 1980]. Two ecologic studies (whole
population) involving 88 Ohio and 64 Louisiana counties suggested
that contaminated surface water was responsible for approximately
8% and 15% respectively of the cancer mortality rate [Page, Harris,
and Epstein, 1976; Harris, Page, and Reiches, 1977]. That's a
large effect, if true.

In its final rule on control of trihalomethanes in drinking
water, EPA [1979a] summarizes many of the risk estimates made for
this carcinogen, and arrives at a lifetime incremental risk of
about 4×10^{-4} of cancer, assuming one drank two liters of water
daily containing 0.10 mg/liter, the newly promulgated maximum level
for this contaminant for community water systems serving more than
10,000 persons. A lifetime risk of 4×10^{-4} seems to be a tolerable
number, at least at first glance. However, this rule does not address the potentially large quantity of chemicals to which one may
be exposed.

How can one get a handle on the potential for chemical waste
contaminating our drinking water and what may be suitable criteria
for their safe disposal? One approach might be to compare the
potential health effects of the annual production of hazardous
chemical waste, and the radioactive waste which would leave a fuel
reprocessing plant annually for each 1000 MWe, light-water-cooled
nuclear reactor, assuming that in each case the waste somehow
found its way quickly into drinking water. A sophisticated calculation of the comparative potential for adverse health effects
would take considerable effort. However, fairly simplistic estimates offer some insight. The arsenic which is being generated in
1980, and presumably will end up largely as hazardous waste, appears to be equivalent in the number of lethal doses, if it were
ingested, to the radioactive waste generated by all currently

operating nuclear reactors (also calculated on an ingestion basis) [Atomic Industrial Forum, 1980; Cohen, 1977].

I asked some of my associates to try to make quick, crude estimates of the comparative hazards of chemical and radioactive wastes. K. A. Solomon used toxicity as a basis. He assumed that he could represent all hazardous chemicals by a sample of ten chemicals. He used Gleason's Clinical Toxicity of Commercial Products to develop an average toxicity for these chemicals and then multiplied by the total amount of hazardous wastes. He then performed a similar calculation to estimate the toxic risk from the radioactive waste produced in one year by a 1000 MWe nuclear reactor. In this way, he estimated that the hazardous chemical waste disposed of each year is about 20,000 times more toxic than the radioactive waste from a single 1000 MWe reactor [Solomon, 1980].

Two other associates, T. McKone and J. Szabo [1980], tried to compare the carcinogenic potential of specific chemical wastes with the radio active waste from a 1000 MWe reactor, assuming ingestion. They calculated a hazard index for each material based on a "consistent" risk value as follows:

$$
HI(m^3) = \begin{cases} \sum_{i=1}^{\text{all nuclides}} \dfrac{Q_i}{MPC_i} & \text{in (curies/m}^3) \qquad \text{nuclear wastes} \\[3em] \dfrac{Q(g)}{WQC} & \text{in (mg/l)} \qquad \text{hazardous chemical waste} \end{cases}
$$

where

HI = Hazard Index for cancer potential (m^3)

Q = Quantity in curies or grams

WQC = Water quality criteria, $\dfrac{mg}{l} = \dfrac{g}{m^3}$, where both the water quality criteria and the radioactive MPC are anticipated to yield a lifetime cancer risk of 10^{-5}.

Alternatively, they also compared the carcinogenic risk using the allowable daily exposure as a measure. They looked specifically at arsenic and the hexavalent chromium ion. In 1977 about 4×10^5 metric tons of chromium were consumed in the U.S. By looking at the kinds of industrial use chromium received, they estimated that perhaps 10% was disposed of the hexavalent form.

In a talk presented late last year at a meeting on risk, which was held at the Nuclear Regulatory Commission, Dr. Elizabeth Anderson of EPA used a viewgraph which presented a summary of water quality criteria based on 10^{-5} cancer risk. The parameter B_H was given for a long list of chemicals including arsenic and chromium, where B_H is an estimate of the lifetime cancer risk in humans from an average daily exposure of 1 mg/day/kg body weight, assuming a linear, non-threshold probabilistic model [Anderson, 1979]. Using a value of $B_H = 43.3$ for chromium, McKone and Szabo estimated that the carcinogenic potential of the hexavalent chromium was roughly equivalent to that of the radioactive waste produced annually by a few hundred, 1000 MWe nuclear reactors. Similarly, assuming an annual disposal of about 10^3 metric tons of arsenic and a B_H of 14 for arsenic, they estimated that the carcinogenic potential of the arsenic was like that of the radioactive waste from two or three, 1000 MWe nuclear reactors. It is my understanding that the carcinogenicity of chromium is subject to considerable controversy. And that the quantitative level of carcinogenicity of arsenic is being re-evaluated. No matter. For purpose of this talk, only orders of magnitude are of interest.

Very elaborate measures are being proposed for the storage and permanent disposal of high level radioactive wastes. To my knowledge, no one has proposed just mixing the radioactive wastes with the chemical wastes and disposing of the mixture in whatever way is currently judged to be acceptable for the chemical wastes alone, taking appropriate precautions against large radiation levels. Why is that? I myself would be unwilling to buy such an approach. But, if so, should I be satisfied with the accepted or proposed practice for disposing of chemical wastes?

EPA has promulgated criteria for classificiation of solid waste disposal facilities and practices under the authority of the Resource Conservation and Recovery Act and the Clean Water Act [EPA, 1979b], and has issued a guidance manual to the states [EPA, 1980b]. In 1979, the EPA issued proposed regulations which identify hazardous waste and provide standards applicable to generators and transporters of hazardous wastes and to owners and operators of hazardous waste treatment, storage and disposal facilities [EPA, 1979c]. In 1980 EPA adopted a set of regulations dealing with the subject [EPA, 1980c].

Since EPA has developed what it considers to be an acceptable approach (at least for the interim) to the future disposal of hazardous chemical wastes, why should I (and you) be loath to assume that the same general criteria are not applicable to high-level radioactive waste (with suitable dilution and other special precautions).

For example, the criteria for hazardous waste disposal sites of the future say that they should not be in a region where they might be flooded by the 100-year flood. This appears to be applied, independent of the content of the disposal site or the possible contamination of potable water, should a greater flood occur. Assuming that thousands of hazardous waste disposal sites will continue to be used, many will be flooded each year. However, the environmental impact statement does not provide a risk-benefit basis for judging that the 100-year flood constitutes a necessary and sufficient criterion for each and every site. If some chemicals are particularly risky, should the criteria be independent of whether large quantities of such a material are present?

Similarily, there exist general criteria for liners and for acceptable soil conditions for a waste disposal site; however, no requirement on liner reliability as a function of time is given. And no basis is provided for judging the degree of confidence required in the ability of the soil to perform its desired function of separating the waste from potable water.

I am not faulting EPA with regard to the proposed criteria. The agency was faced with a very difficult job. The generation of hazardous waste is intimately connected with a large fraction of modern-day life.

The past practice for handling such waste was inadequate, and there was need to develop some approach on an expedited time scale.

However, now that an approach has been formulated, at least in part, it appears appropriate to apply risk-benefit analysis to the problem in a host of different ways. Not that one can get good answers to many of the questions. Not that one can expect agreement between such analyses when performed by separate groups, especially when they have differing motivations.

Nevertheless, the process of trying to perform such analyses and to evaluate them should illuminate areas wherein knowledge is grossly insufficient; it should raise questions on the necessity and sufficiency of specific regulations; and it should provide a more structured approach to attacking the entire problem.

REFERENCES

Anderson, E., Presentation at Subcommittee Meeting of Advisory
     Committee on Reactor Safeguards, December 1979, U.S. Nuclear
     Regulatory Commission.

Atomic Industrial Forum, 1980, Statement of position on the storage and disposal of nuclear waste in the matter of Waste Confidence Rulemaking of the U.S. Nuclear Regulatory Commission, July 7, 1980.

Cohen, B. L., High-level waste from light water reactors, 1977, Rev. Mod. Phys., 49(1):1-20.

Environmental Protection Agency, 1979a, National Interim Primary Drinking Water Regulations; Control of Trihalomethanes in Drinking Water; Final Rule. Federal Register, 44, No. 231, November 29, 1979.

Environmental Protection Agency, 1979b, Criteria for Classification of Solid Waste Disposal Facilities and Practice. Federal Register, 44, 53438, September 10, 1979.

Environmental Protection Agency, 1979c, Draft Environmental Impact Statement for Subtitle C, Resource Conservation and Recovery Act of 1976, January 1979.

Environmental Protection Agency, 1980a, "Planning Workshops to Develop Recommendations for a Ground Water Protection Strategy."

Environmental Protection Agency, 1980b, "Classifying Solid Waste Disposal Facilities: A Guidance Manual," EPA Report SW-828, March 1980.

Environmental Protection Agency, 1980c, Federal Register, 45, 33063, May 19, 1980.

Harris, R. H., Page, T., and Reiches, N. A., 1977, Carcinogenic hazards of organic chemicals in drinking water, in: Origins of Human Cancer, H. D. Hiatt, J. D. Watson, and J. A. Winsten, eds., Cold Spring Harbor Laboratory, Cold Spring Harbor, N.Y.

Kimm, V. J., Kuzmack, A. M., and Schnare, D. W., 1980, "The Questionable Value of Cost-Benefit Analysis: The Case of Organic Chemicals in Drinking Water," Environmental Protection Agency, March 11, 1980.

McKone, T. E., and Szabo, J., (1980), unpublished, UCLA, Los Angeles.

Page, T., Harris, R. H., and Epstein, S. S., 1976, Drinking water and cancer mortality in Louisiana, Science, 193:55-77.

Solomon, K. A., (1980), unpublished, UCLA, Los Angeles.

METHODOLOGY AND MYTH

William D. Rowe

Institute for Risk Analysis
The American University
Washington, D. C.

INTRODUCTION

Risk assessment is now new. Reliability and quality control
techniques, the science of gambling, and economic risk analysis have
been with us for decades. In general, all of these approaches have
dealt with common events for which historical data may be reasonably
required and the consequences can be objectively measured. Value
judgments about gambles are left for managers or gamblers themselves.

In the last few years, attempts have been made to extend the
methodologies developed for common events to the more difficult problem of risk assessment of broader societal problems. Many methods
have been put forth and attempted with varying degrees of success.
Moreover, considerable attention is directed at the development of
universal methods for risk assessment. Unfortunately, many such
methods are like the Holy Grail, an unattainable, mythological goal.

The extension of risk assessment to societal problems immediately runs into two fundamental classes of methodological problems.

The first class involves the estimation of the probability of
occurrence of rare events. Ordinary events occur often enough so
that the central limit theorem applies. This is not true for rare
events, of which there are, at least, two kinds: (1) the zero-
infinity dilemma of a low-probability high-consequence event and
(2) the signal imbedded in noise, such as the detection of cancer
causes in a high level of spontaneous cancer occurring normally.
These are primarily problems of measurement, and there are definable
limits to knowledge, both in the practical sense and in theory.

The second class of problems involves the value judgments that must be addressed when trading off the benefits of new or existing activities against the residual level of risks inequitably borne by parts of the population. Basically, this is a question of a level of acceptable risk being established, either formally or informally. These are primarily problems of social behavior and values where wide variations are the rule.

This paper will address these problems and the limitations of knowledge and methods in further detail.

RISK AND RISK ASSESSMENT

The scientific study of risk and the technology of control are not new. They have been a part of the knowledge base for years. Gambling games, formal and informal, are as old as man, just as the risks of doing business have been recognized for years with a variety of formal structures provided for estimating and evaluating such risks. Reliability and quality control techniques have progressed to very sophisticated levels. Moreover, designing for safety now is part of engineering practice, and safety engineering has even become a discipline. Statistical methods have provided epidemiology, toxicology, and ecology with a wide variety of useful tools. However, these approaches have evidently not had much effect in ameliorating public concern with risks. Why they have not been effective, along with questions of what is risk, what is risk assessment, what are acceptable levels of risk and how are these levels controlled are all valid questions today.

Risk assessment is a tool for dealing with such problems, but as a tool it can only provide aids to solutions--it is not a solution itself. The study of risk assessment can broaden our understanding of the problem of dealing with risks in the business enterprise and larger society. Risk assessment is not a substitute for enlightened planning and decision making. Analytical techniques must be used with common sense, especially when dealing with value problems and associated human judgments that must be addressed. Even with these limitations, risk assessment can help focus our thinking on the critical issues and parameters and make implicit value judgments visible.

What is Risk

Risk is the chance of harm. Mankind has always been subject to risk, and will continue to be. The concern of an industrial manager is to balance a variety of risks--technological, economic, and personnel--in the pursuit of organizational goals. These are

risks which the organization voluntarily adopts as the risks of conducting a business. It is only when such risks are undertaken with
lack of knowledge about the full scope of risks, or when these risks
are in conflict with societal concerns, that problems develop in risk
management.* In contrast, the concern in society today is focused
primarily on involuntary risks imposed upon members of society by
technology whereby the risk takers do not necessarily share in the
benefits of the technology.

Definitions

More formally, risk is the potential realization of unwanted
consequences of an event. Both the probability of occurrence of an
event and the magnitude of its consequence are involved. The term
hazard implies the existence of some threat, whereas risk implies
both the existence of a threat and its potential for occurrence.
Thus a threat (hazard) may exist without implying risk. Risk occurs
if a potential pathway for exposure exists, i.e., there must be some
exposure pathway to man, biota, or the environment for a threat to
be meaningful. In this sense, the level of exposure can be related
to the likelihood (probability) of occurrence and the magnitude of
consequences of an event.

Risk, then, is some function of probability and the value of
the consequence determination:

$$Risk(R) = f(p, C_{(v)})$$

where p is the probability of occurrence and C(v) the value of
the consequence. When the function is multiplicative, the result
is called expected value of the risk.

The very definition of risk as given above is inadequate when
considering societal risk holistically. It focuses only on the negative side and tends to foster unwarranted concern on risks.

A more general definition of risk is "the downside of a gamble."
While not in conflict with earlier definitions, it broadens the concept to require that a gamble, with gains and losses, be undertaken
in order to have risk. Living, itself, involves gambles--some of
which are involuntary, but which often require trade-offs between
the quality of life and its quantity (longevity). In this light,
the major concerns of risk assessment are focused on inequitable

---

*The term risk management is used to address these organizational-
type risks, while the term risk assessment is used to encompass the
broader societal aspects as well.

gambles, i.e., a gamble where one part of society gains while
another part takes the risks.

When specific inequitable risks are addressed, they are usually
elevated in relation to the specific problems addressed. Concern
for the specific risks in question often dominate the analysis.
Gains, the upside of the gamble, are sometimes considered in terms
of cost-benefit analysis, but the attention to benefits is usually
localized to the specific problem.

In a general sense, every technological undertaking (and many
social ones) involves a gamble, with some residue of inequitably
imposed risk on parts of the population. In this light, there is
no such thing as zero risk, only involuntary and voluntary gambles
for which minimum risk for acceptable gain is one fuzzy criterion
for decision making. Given this perspective, the problem of estimating risks and evaluating them can be broadened.

As stated previously, risk assessment and analysis is not new--
it has been with us for years. What has changed in the determination of risk is a new focus on rare events without historical precedence and an increased magnitude and variability of even consequences.
In both cases, there are quantifiable limits to knowledge which limit
the ability of science to contribute to technical solutions. Two
kinds of rare events, the "zero-infinity dilemma" and the "masked
even" case, will be discussed in subsequent sections of this paper.
Recognition of the limits to knowledge can provide the impetus for
strategies to make decisions in spite of uncertainties.

For the social evaluation of risk the growth of adversary
approaches to decision making has led to polarization instead of
ameliorization. More importantly, value judgments cannot remain
hidden if credible solutions are sought. On the other hand, acceptable levels of risk exist. The challenge is to develop processes
whereby all risks can be eventually put into perspective in acceptable societal gambles.

The Risk Assessment Process

An initial classification of risk assessment can provide a
first cut of a reasonable system for grouping the types of problems
involved and approaches to solution. The term "risk assessment" is
used here to describe the total process of risk analysis, which
embraces both the determination of levels of risk and social evaluation of risks. Risk determination, in turn, consists of both identifying risks and estimating the likelihood and magnitude of their
occurrence. Risk evaluation measures both risk acceptance (acceptable levels of societal risk) and risk aversion (methods of avoiding

risk, as alternatives to involuntarily imposed risks). The relationship among the various aspects of risk assessment is illustrated in
Figure 1.

OBJECTIVE AND SUBJECTIVE RISK ESTIMATION

Expected risk, however, cannot be fully equated with actual risk
because probability and consequence estimates may be inexact. The
probability that a consequence will occur may be determined by direct
measurement of repeated trials of a causative event. When the number
of trials is large, the estimate of probability is properly considered objective, because it represents an empirical estimation. On
the other hand, if probability estimates are made on the basis of one
or a few trials or by conjecture, then the probability estimate is
subjective. These definitions comport with classical definitions of
objective and subjective probability.* Between these extremes lies
another estimate, here termed "synthesized probability," for which
the probability of a consequence is not measured directly, but rather
is extrapolated from the objective probabilities of causative systems
that are expected to behave similarly. For example, the Nuclear Regulatory Commission's study on reactor safety** uses an estimate of
probabilities computed from tests on parts of the reactor system and
synthesized into a total model.

A similar range of certainties exists in calculating consequence
value. An objective estimate of consequence value consists of a
consequence that is directly observable and measurable and a consequence value that is expressed explicitly. For example, the accounting value of the pay offs of a gambling establishment represents an
objective consequence value. Subjective consequence value, on the
other hand, occurs when the consequence value to a particular risk
agent depends completely on the risk agent's personal value system
and situation. Between these two extremes is a value estimate, here
termed "observable consequence value," in which the behavioral

---

*Subjective probability is a number between 0 and 1 assigned to an
event, based on personal views whether the event will occur. Objective probability is a number between 0 and 1 assigned to an event,
based on a history of trials of similar events estimating whether
the event will occur. See generally "Dictionary of Scientific and
Technical Terms" 1974, 10:720-726. Concerning subjective estimates,
the Bayesian approach to conditional probability and the use of
a priori information is useful. See R. Schlaiffer, 1964, "Analysis
of Decisions Under Uncertainty."

**U.S. Nuclear Regulatory Commission, 1975, WASH-1400, "Reactor
Safety Study."

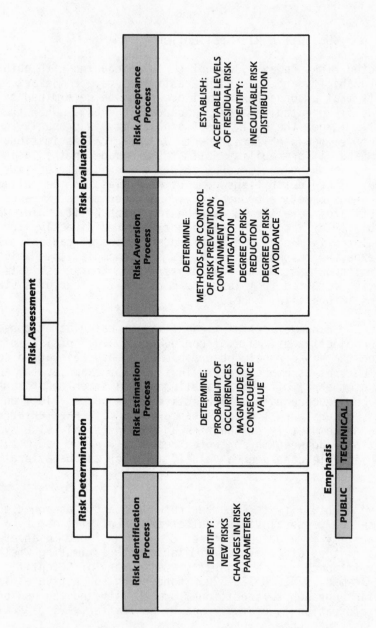

Fig. 1. The components of risk assessment

response of groups in society to objective or subjective risk consequences is ascertained and measured by studying actual behavior.

A relationship between the nature of probability and the nature
of consequence, both of which range from the objective to the subjective, is illustrated in Figure 2. Probability is diagrammed
vertically and consequences horizontally. The combination of probability and consequence defines risk. Thus, the combination of
objective probability and objective consequence defines objective
risk. Most scientific studies of risk have concentrated on objective risk because such risks are easiest to define and measure. As
the diagram moves towards synthesized probability or observable consequences, it defines an area termed "modeled risk." Such risks are
not directly observable or objective, and the usefulness of computing modeled risk depends on the degree to which the model corresponds
to reality. Risk may be modeled by using synthesized probability,
or observable consequences, or both. All other risks are subjective
because they are computed on the basis of subjective estimates, subjective valuations, or both.

The science of risk estimation traditionally used scientific
experiments and empirical measurements to compute objective risk.
Recently, the concept of synthesized probability has developed
extensively, just as measurement of observable consequences has
gained major importance in the behavioral sciences. By contrast,
societal decision making relies on subjective risk estimates; in
practice, emotional considerations are more compelling than objective scientific knowledge. Objective estimates of the risks of
nuclear reactors, for example, are usually lower than public perception of them,* yet public perception may have greater impact on
governmental decisions whether and how to utilize this form of fuel
production than scientific estimates. Because objective risk estimates provide a sounder basis for industry and governmental decision
making, administrators could try to persuade the public to accept
such measurements. This can be accomplished either by educating the
public to understand objective risk or by making subjective risk
estimates and valuations more explicit and visible so that the public understands their fallibility.

SUBJECTIVE NATURE OF RISK ESTIMATION

The following questionnaire is used to sensitize the reader to
the subjective nature and meaning of risk estimation problems. Four

---

*Kates, 1976, Risk assessment of environmental hazard, Scope, 8,
Report 151, publication of the International Council of Scientific
Unions, Scientific Committee on Problems of the Environment.

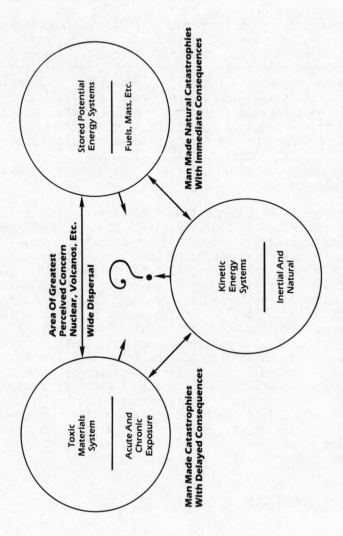

Fig. 2.   Systems prone to catastrophic events

questions are posed, with an answer sheet provided in Table 1. The
reader is asked to put himself in the role of a responsible decision
maker who is required to make a decision, as unattractive as the
decision may be to make. The reader is to use his own values and
reasoning as to how he would make a decision. The questions illustrate, in a stripped-down manner, the kinds of questions risk analysts ask of decision makers. The analyst ought to understand his
own values, however, before asking decision makers to state their
values.

Questionnaire

Question 1. You are faced with two and only two alternatives:

Alternative A--During the next year one (1) person will be
killed with certainty as a result of this selection.

Alternative B--During the next year there is a probability, P,
of .01 that one hundred (100) people will be killed in one incident as as a result of this selection.

Discussion. Alternative A is a certainty of one death per year.
Alternative B is the expected value of one death per year, i.e.,

$$E(v) = .01 \times 100 = 1$$

If you are indifferent to the two alternatives, i.e., certainty is
equated with expected value in this case, you would be considered
risk neutral. If you feel that the chance of 99 in 100 of nothing
happening next year is preferable to a certainty of one fatality,
although there is a slight chance (.01) of 100 fatalities, then you
would be considered as risk taking, and you would prefer alternative
B. Conversely, if you feel the slight chance (.01) of a catastrophe
(100 fatalities) occurring outweighs the certainty of one fatality,
you would be considered risk averse, and you would prefer alternative A. There is no right answer, only an answer that expresses
your own valuation. On this basis the following is required for
answers (Table 1):

Requirement

a. Choose A or B or indifference;

b. Select a value for P on the scale of Table 2 for which
    you would be approximately indifferent between the two
    alternatives.

The second requirement requires consistency with the first,
i.e. P = .01 for indifference, .01 < p < 1 for risk taking, and

Table 1. Value Query Tally Sheet

Date			Name (optional)
Case 1	a.	A B	(Circle choice or both for indifference)
	b.	p = _____	(Enter value)
Case 2	a.	A B	(Circle choice or both for indifference)
	b.	p = _____	(Enter value)
Case 3	a.	A B	(Circle choice or both for indifference)
	b.	n = _____	(Enter value)
Nuclear Power		For Against No position	(Circle choice)
Case 4	a.	A B	(Circle choice or both for indifference)
	b.	% _____	(Enter value)

Table 2. Probability Scale (Logarithmic)

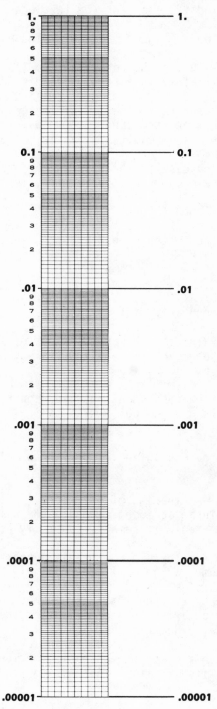

p < .01 for risk aversion. The product of P and 100 is the
expected value of indifference for you (excludes equality).

Question 2. You are again faced with two alternatives as before:

Alternative A--During the next year one (1) person will be
killed with certainty as a result of this selection.

Alternative B--During the next year there is a possibility,
P, of .001 that one thousand (1,000) people will be killed
in one incident as a result of this selection.

Discussion. The case is the same as the previous case, except
that the expected value, which is still one, is calculated by:

E(v) = .001 × 1,000 = 1

Does a larger, but less probable consequence cause a change
from the previous case?

Requirement

a. Choose A or B or indifference;

b. Select a value for P on the scale of Table 2 for which
    you would be approximately indifferent between the two
    alternatives.

You must again be consistent between requirements a and b.
Here indifference is a P = .001; risk taking, .001 < p < 1; risk
aversion, p < .001.

Question 3. You are again faced with two alternatives, and
only two:

Alternative A--At the end of N years (in this case N = 5)
you will die in an accident although you will not anticipate
the accident. You will remain reasonably healthy in the interim.

Alternative B--At the end of three (3) years you will experience
a malignant tumor which will progressively get worse until you
die at the end of 10 years. Your health will slowly diminish
over this time until the 10th year when it will progressively
worsen, causing terminal hospitalization, dehabilitation, acute
pain, and finally death.

Discussion. This question contrasts the length of life (quantity) with the quality of life. There is no right answer, only an

expression of your own value system. There are no alternatives that you should infer above those shown.

Requirements

a. Choose A or B or indifference;

b. At what value of N (years) would you be approximately indifferent between A and B?

Again, b. must be consistent to a. If you are indifferent, N = 5. If you value the quality of life over quantity (alternative A) then 0 < N < 5. The value of N may only be less than N = 3 if you fear cancer so much that you would sacrifice some years of quality life to avoid cancer. Conversely, 5 < N < 10 if you value the length of life over the quality, alternative B.

Question 4. Before addressing question 4, ask youself whether you consider yourself pro-nuclear, anti-nuclear, or haven't made up your mind when it comes to nuclear power plants. Enter this on Table 1. Now choose between two alternatives:

Alternative A--In the next 10 years a nuclear meltdown occurs with 10,000 people killed; however, energy independence results.

Alternative B--In the next 10 years there is a 10 percent chance of world war as a result of U.S. dependence on OPEC oil.

Discussion. The perspective of a total gamble can affect one's attitude toward risk. It is assumed in alternative A that nuclear power is risky, but such risks might be warranted if a war can be avoided.

Requirements

a. Choose A or B or indifference;

b. At what percent of possible world war would you be indifferent?

Consistency between a and b requires that the percent chance of war be 10% for indifference; 10% value < 100% if you selected alternative B, i.e., how imminent would war have to be before you were indifferent between A and B; 0 < value < 10% for alternative A.

Discussion of the Questionnaire Results

Over 200 people have been subjected to this questionnaire at this writing, including risk analysts, Environmental Protection Agency staff and contractors, students, and others. The results reflect the wide variety of values.

Questions 1 and 2. The results show that a few people (about 20%) are risk neutral for either question 1 or 2. A bit less than 50% selected alternative B in question 1 and a bit less than 40% selected alternative B in question 2, i.e., they are risk takers. The remainder are risk averse. About 70% did not change their view as the size of the consequence increased.

The most important result is quantitative. No one has been more than two orders of magnitude risk averse, i.e., no one had a value of P less than .0001 for question 1 or a value of P less than .00001 for question 2. While the sample and experiment is limited, it provides an important indication that people generally have practical limits on risk aversion. This range of two orders of magnitude for risk aversion will be used subsequently for elimi- nating alternative risks from consideration in some cases.

Question 3. About 80% of those polled chose alternative A, i.e., they favored a shorter disease-free life. Less than 3% were indifferent, but about 18% preferred a longer life. The important points are twofold: (1) the value responses are distinctly bimodal since there are those who value quality and those who value quantity predominantly; (2) the majority selected quality over quantity. These are important points for risk evaluation. In the first case any decision involving trade-offs between quality and quantity of life will seldom be unanimous, there will always be valid, opposing value positions. In the second case any decision that only focuses on the length of life without regard to quality may meet stiff oppo- sition. A good example of the first case are people on both sides of the nuclear power problem. An example of the implication of the latter case was the public and congressional opposition to the FDA on its decision to ban saccharin.

Question 4. When one views the total gamble, decisions often become clearer. In one study, 47% declared themselves pro-nuclear, 23% anti-nuclear, and 31% had no position prior to addressing ques- tion 4. Afterward, 59% selected alternative A, 31% selected alter- native B, and only 10% were undecided. The number of undecided was reduced in terms of the specific gamble. Over 90% would have selected A if the percentage of war was 50%, i.e., unsafe nuclear power is preferred to a one-in-two chance of war based upon achiev- ing or not achieving energy independence.

The purpose of the questions is to show how difficult value judgments are when stripped bare, that there are no right answers, that the range of values is diverse and often opposing, and that there are practical aspects to dealing with such issues. The observed limits of risk aversion and the change in values as the total gamble is presented are cases in point. These can be applied to rare events.

ORDINARY AND RARE EVENTS

Rare events are considered to be those which occur so infrequently that direct observation is improbable. These events are particularly important in two different aspects of risk estimation. The first aspect is the "zero-infinity dilemma" [Page, 1979] for risks with very low probability and very high consequences such as those associated with nuclear power plant meltdowns. The second is the determination of risk where there is a very low probability of occurrence and the numbers of people exposed are high, but where the measurements of effects are masked by spontaneous occurrences, uncontrolled variables, conflicting risks, synergistic and antagonistic processes with other threats, etc. These kinds of events occur when we attempt to detect cancer in animals and human populations caused by substances whose potency is not very high. The first type of rare event involves the problem of accidental exposure to toxic chemicals; the second to the problem of potency measurements.

Definitions

A rare event may be defined as:

$np < .01/y$, or

$np < NP$

where n is the number of trials, p the probability of event occurrence in a test population, N is the total number of trials occurring in the parent population, and P is the number of spontaneous and competing events occurring in the parent population. Both of the definitions above are arbitrary. The first refers to the zero-infinity dilemma and our inability to acquire historical data for something that occurs less often than once in a hundred years. Other values may be used. The second is the problem in measuring potency (the problem of measuring a signal imbedded in noise). An event which is not rare we will term as ordinary event.

The definitions are based upon Bernoulli trials and hold equally well for binomial and Poisson processes. It is important to note that when $np \geq 5$ the binomial and Poisson distributions are closely approximated by a normal distribution. Below this value, increasing error occurs as np becomes smaller. Thus, there are two classifications, as shown in Figure 3, dealing with rare/ordinary and normal/non-normal events. Of course, the factor NP is shown arbitrarily in the case np(2) in Figure 3 and can vary over the entire range of np, including levels higher than 5. In this latter case, statistics based upon normal distributions may be used. These may include analysis of variance, regression analysis, and multiple correlation techniques.

Probability and Belief

All probability estimates involve degrees of belief and are by definition subjective. The three classical approaches to estimating probability are given below in increasing order in terms of the degrees of belief required in each case.

a priori information--prior knowledge about the behavior of a system for which one has a degree of belief that similar behavior can be expected to occur in the future, e.g., knowing in advance that a particular coin toss is "fair."

likelihood of occurrences--study of historic or experimental data to determine the behavior of a system in order to evaluate its future behavior, e.g., observing the outcomes of a roulette wheel to determine possible imbalance. Here there is a degree of belief about the validity of the experiments as well as for the continuance of similar behavior.

subjective estimates--in the absence of historical data, the use of any available information to estimate probabilities and subjectively evaluate the meaningfulness of the information used, e.g., betting on a particular football game. Here the degree of belief involves the validity of available information as well as the kinds of degrees or belief involved in the two cases above.

For rare events it has been traditional to use a combination of these approaches to form two other approaches:

modeled estimates--a study of the behavior of similar systems for which data is available, which, with reasoned modification, is used as a model for the system under analysis, e.g., the estimate of rupture of steam boilers in general to provide an estimate of the probability of rupture of nuclear reactor boilers. Here the belief structure involves the confidence one has in comparing such systems, e.g., does radiation damage increase failures in boilers?

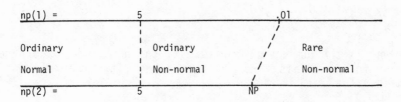

Fig. 3. Classification of tests by the product np

system structuring--the failure of systems may be rare because
of redundancy, so analysis of the failure probability of component parts and their interconnection is used to synthesize
an estimate of system behavior, e.g., event trees and fault
trees in nuclear reactors. The belief structures involve the
degree of knowledge about individual component behavior, how
components behave in a system, and the degree to which all
important system combinations can be ascertained. Such systems
are always open-ended since the combinational possibilities
are astronomical in number.

Realizing the limitations of these approaches, the concept of
putting error ranges on the estimates of probability has evolved;
such error ranges were used in the case of the Reactor Safety Study
conducted by Norman Rothman for the Nuclear Regulatory Commission.
The object is to estimate the confidence one has in his estimate.
This represents a "degree of belief" about one's "degree of belief."
Without a keen understanding of the implications and limitations of
such approaches, they have often been applied erroneously, as was
the case for the Reactor Safety Study as pointed out by a subsequent
review committee [Lewis Committee Report, 1978]. Even worse, many,
including the media and the public, have confused an estimate of the
probability of occurrence of an event with a prediction of future
action, i.e., clairvoyance. Statisticians and scientists know that
probabilities of future outcomes are but measures of the relative
likelihood of the alternatives.

Rare Events

Many rare events are considered to obey a Bernoulli process,
i.e., they are independent of each other and occur with a probability, p, or do not occur with a probability 1 - p. At the simplest
level this process is described by a binomial distribution. The

binomial distribution is of the form:

$$p_x(x) = \binom{n}{x} p^x (1-p)^x \qquad x = 0,1,2,\ldots,n$$

where n is the number of trials, x the number of occurrence of
an event in n trials, and p the probability of an event occurring.
The mean value is:

$$\bar{x} = np$$

and the variance is:

$$\sigma^2 = np(1-p)$$

For the case of p = 5, this is equivalent to flipping a fair coin.
For rare events very small values of p are of interest (large values
of 1 - p for nonoccurrence). For small values of p, the variance
is approximated by:

$$\sigma^2 \approx np$$

The coefficient of variation is a measure of the degree of uncertainty
and is defined as the standard deviation divided by the mean.

$$C.V. = \frac{\sigma}{\bar{x}} = \frac{np}{np}$$

    The significance is that np is a small number for rare events
and the square root of a small number is always a larger number. The
form of the distribution of uncertainty is shown in Figure 4 for
n = 1. The coefficient of variation is inverted at p = 0.5 for
symmetry. On the left, the coefficient of variation tending toward
zero is a measure of certainty. On the right, the inverse is a measure of uncertainty. At a value of p = .1 the standard deviation
is three times the mean value, already a very uncertain situation.
As n is increased, the C.V. only improves by the square root of
n. This same curve holds for the Poisson distribution.

    The important point missed by many investigators is that the
very nature of the distribution indicates wide ranges of uncertainty
for values of p which are very small. Thus, expected value becomes
most meaningful at values of p = 0.5 since both p and (1-p) are
of their maximum values. Figure 5 illustrates this point by indicating that the meaning of expected value falls off toward zero in terms
of absolute risk on either extreme of the probability scale shown as
the abscissa. The left-hand abscissa refers to the meaning of the
expected value. The right-hand abscissa indicates that, as the meaning of absolute risk decreases, the meaning of relative risk becomes

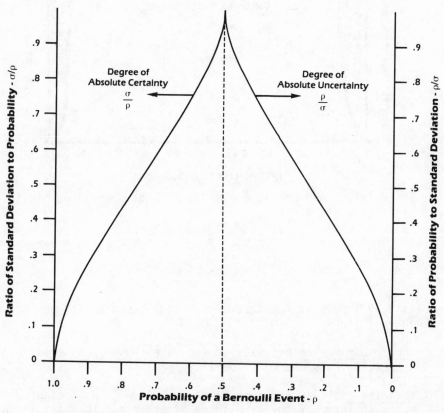

Fig. 4. Variability of knowledge of a Bernoulli event as a function of its probability

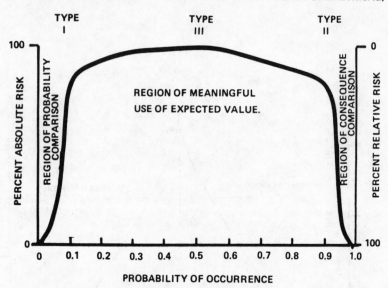

Fig. 5.   Relative and absolute risk as a function of probability of
consequences

increasingly important. Relative risk and absolute risk may be
defined as follows:

absolute risk--an estimate of the likelihood of an event with
a specific consequence (Type III).

relative risk--an estimate of the relative likelihood of an
event in terms of the likelihood of other events of a similar
magnitude (Type I) or the comparison of event magnitudes for
events with the same likelihood (Type II).

The Absence of Information as Information

The absence of occurrence of events is in itself information.
Two cases can be illustrated: (1) the limits of risk for zero
events in a set of independent trials and (2) an estimation of the
mean time to failure for a system from the number of years of
failure-free operation. The first case is a Bernoulli process and
involves calculating the probability of a Bernoulli event, p,

from the P_0 term of a binomial distribution for N such events:

$$P_0 = (1 - p)^N$$

To estimate p to level of confidence, γ, (e.g., $\gamma = .99$, where $\gamma = 1 - \alpha$) we use:

$$\gamma = 1 - (1 - p_\gamma)^N$$

i.e., $p < p$ with a confidence of γ.

$$P_\gamma = 1 - (1 - \gamma)^{-N}$$

The second case involves a Poisson process so that if t is total number of system years of operation and failures occur randomly in time, it is possible to deduce an upper limit on the probability of failure per year. Such a limit is longer than or equal to the failure rate, λ, with a given confidence level ($\gamma = 1 - \alpha$). If $N(t)$ is the number of failures up to t system years, the upper confidence boundary is

$$\gamma_1 = \chi_\alpha^2 / 2t$$

where χ_α^2 is the chi square distribution with $2(N(t) + 1)$ degrees of freedom [Thedeen, 1979].

For identical multiple systems the system years are the sum of the years of operation of all systems.

The first case can provide an upper limit on the risk of cancer in animal tests which have negative results (no excess cancers). The second case can provide an upper limit on risk for cases where systems are operating for t system-years without an event of the type being considered occurring, i.e., $N(t) = 0$. In this case, systems must be identical to use systems-years in a valid manner. For example, this approach has been attempted for nuclear reactor melt-downs by using the operating history of all reactors, civilian and military, light- and heavy-water reactors, etc. The homogeneity of reactor types for meltdowns has never been established.

ABSOLUTE RISK VS. RELATIVE RISK

For a go/no-go type of decision, one would like to have a mean-ingful absolute risk estimate. For selection of one of a set of alternatives, only relative risk estimates are required. As will

be seen, relative risk evaluations can be quite useful in decision
making.

Absolute risk estimates may or may not be useful for decision
making, depending upon where the risk estimates and their ranges of
uncertainty lie. Decisions are always made against some reference
or set of references. Benchmarks are one form of reference that do
not necessarily imply acceptability. They are risks of a similar
nature that people have experienced, and they provide a reference
to real conditions. However, if the results of analysis show that
the range of uncertainty in estimates of probability of occurrence
encompass reasonable benchmarks, resolution of the decision by probabilistic methods is unlikely. If the benchmarks fall outside the
range of uncertainty, then probabilistic approaches can be effective.

As an example, the worst estimate of risk for high-level radioactive waste disposal seems to be lower than benchmarks that seem
to be in the acceptable range [Martin, 1979]. If the bands of uncertainty of probability estimates encompass the range of acceptable
risk levels, the decision cannot meaningfully be based on probability
estimates. Figure 6 illustrates this problem using nuclear accidents
and high-level radioactive waste disposal as examples. The scale on
the left is a measure of absolute risk in terms of the probability
of the number of fatalities that might occur in a year. Some benchmarks are shown on the right, including worldwide fallout from nuclear
weapons already committed, planned releases from the nuclear fuel
cycle for 10,000 Gwe-years of operation (the maximum production possible from available uranium resources without breeding), one percent
of natural radiation background, and radon and radiation from undis-
turbed uranium ore bodies. These benchmarks are only to provide
perspective; they do not, by themselves, imply acceptability.

The ranges of risk estimates for a high-level waste repository
for all high-level wastes (10,000 Gwe-years) lies well below the
benchmarks. Thus, a decision on high-level waste is resolvable by
probabilistic methods. The range of risk estimates for all nuclear
reactor accidents (10,000 Gwe-years of operation) is shown based
upon WASH-1400, WASH-1400 COMMENTS, the Lewis Committee report, and
extrapolation. The exact range may be argued, but it probably en-
velopes all of the benchmarks, making any decision based upon probabilistic analysis alone indecisive. In this case, although it may
be possible to refine the estimates to some extent, it may be impossible to reduce the residual uncertainties to a level for which
meaningful decisions can ever be made by this approach. Thus, some
decisions may be truly indecisive by this approach, while others
can be decided. Indecisive refers only to a probabilistic solution,
since many other approaches, including social-political analysis,
may still be effective.

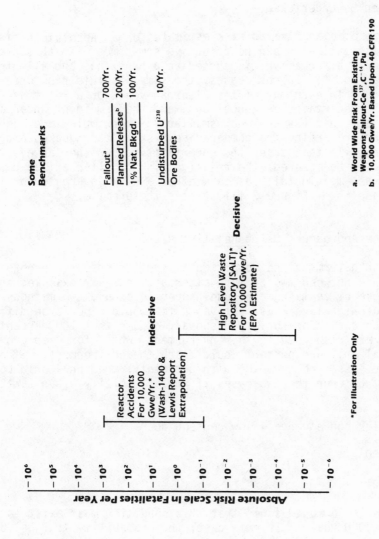

Fig. 6. Absolute risks, uncertainty, and benchmarks

A reasonable management strategy for investigating the probability of rare events is to proceed in two steps. The first would be a preliminary analysis to determine the range of uncertainty bands in probability estimations, which then would be compared to benchmarks to determine whether overlap exists or not and to what extent. Then, the value of information provided by a second, more-detailed analysis can be determined. If warranted, the more-detailed analysis can then be undertaken.

In the absence of valid risk estimates on an absolute basis, relative risk estimates can be effective sometimes. In this case, one of several alternatives is chosen as a baseline, and all others are compared to it. In most cases, the same absolute risk uncertainties occur in each one of the alternatives. On a relative basis, these uncertainties cancel out, and the remaining uncertainties among alternatives are much smaller. For example, one might ask questions regarding the comparative risk of earthquakes among alternatives. In this case, the uncertainty of whether an event will occur may be ignored; the relative risk of alternatives can be established meaningfully as to whether a risk is greater or lower than the baseline. Figure 7 illustrates the premise.

Indifference Approaches Using Relative Risk

One can determine the condition of indifference for probability of occurrence between two or more alternatives. For example, should an earthquake of certain magnitude happen to occur, two methods of disposing of high-level radioactive wastes would experience different consequences. For surface disposal, assume that 100 percent of the wastes are dispersed to the environment, while for deep geological disposal only one percent would reach the environment. If one assumes the lesser consequence with certainty, then one would be indifferent between the two cases if the probability of an earthquake was .01/yr.

p × high consequence = low consequence

p × 100 = 1

p = .01/yr.

However, one must assume that this condition must exist for, perhaps, 10,000 years. In this case, the probability must be .01 for 10,000 years or 10^{-6}/yr. One then looks at the absolute probability estimates for earthquakes in the region of concern. These estimates can come from models, the subjective judgment of experts, or can be reserved for the decision maker and his expectation of the future. If the estimates or the decision-maker's expectation are

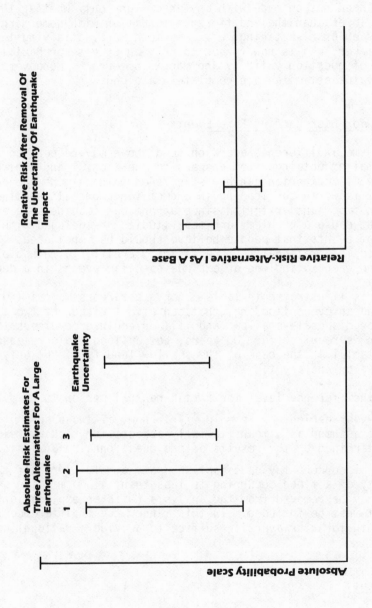

Fig. 7. Absolute risk to relative risk

higher, deep geologic disposal with the smaller consequence is chosen, and vice versa, assuming the decision maker is risk neutral.

The important aspect of this approach is that absolute risk estimates need not be made with any precision--only whether they are higher or lower than the indifference condition with some margin of risk aversion or risk taking. This is much more satisfying to a decision maker, who is now forced to rely on precise probability estimates of question validity and margin of error. Moreover, expected value estimates are completely avoided.

Criterion for Disregarding Rare Events

The same indifference approach used above may be coupled with the information obtained from answers to questions 1 and 2 given previously. A criterion can be established to determine whether a rare event is worth considering in a decision among alternatives. This is an important criterion since anyone can imagine some event which might cause a problem, and most studies are deluged with a plethora of events that people believe should be considered. What is required is a rational (but necessarily arbitrary) method of determining when to include or exclude specific events in a decision.

This is illustrated in Table 3 for a hazardous-material disposal problem and an event flooding, that can affect either of two alternatives for disposal--shallow land fill or engineered disposal crypt. Protection is required for 100 years, but a flood would release 100% of the material in the case of the shallow land fill, but only 1% in the case of the crypt.

The indifference level for a risk-neutral person is $p = .01$ for a 100-year period, or $p = 10^{-4}$/yr. Now, if two orders of magnitude are assumed as a means to fully account for risk aversion, as was determined by the results of the questionnaire analysis, a criterion of concern may be established at $p = 10^{-6}$/yr. If the probability of a flood occurring is less than 1 in a million per year, it may be removed from the analysis (with an explanation); if not, it must be included. In this manner the indifference level and a margin for risk aversion suffice to provide a rational criterion.

Masked Events

Rare events that are rare because they are masked by other, more numerous, competing events are another type of problem. These kinds of events occur, for example, when one attempts to detect cancer in

Table 3. Hazardous-Waste Disposal, Hypothetical Example

Parameter	Disposal Method	
	A Shallow Land Fill	B Engineered Disposal Crypt
Event (Flooding)	100% of Material Dispersed	1% of Material Dispersed
Risk Neutrality Condition	$\rho = .01$	$\rho = 1$
Postulated Containment Period	100 Years	

animal or human populations caused by substances whose potency is not very high. The discussion on this aspect of rare events is important and extensive. It is not covered here due to space limitations.

VALUE DIVERSITY PROBLEMS

Question 3 demonstrates the kind of diverse values that exist among groups of people. Such values can stem from pre-examined positions, belief structures, or personal preference. There is no absolute right or wrong, only different sets of values. One set of values viewed from another point of view may seem wrong, but what is the "standard" value? The resolution of multivalued positions is a political problem, but the analyst may legitimately ask two questions: (1) how can I present the decision so that value differences are made visible and their impact understood? and (2) how can I present the analysis in a manner that aids in value convergence as opposed to causing divergence?

Visibility of Value Judgments

The first question addresses the need to make visible the value judgments and the main arguments on various sides of the judgments. Methods that apply weight to each value and attempt to provide single scales of judgment tend to mask such judgments. They are made at lower levels by analysts with their own values implied, and attempts

to extract them by consensus techniques (e.g., Delhi) end up satisfying few. If an analyst expects a decision maker(s) to provide his
value judgments a priori, he will be disappointed. Expressing numerical values a posteriori is difficult enough, as demonstrated in the
four questions posed earlier. It is impolitic for a decision
maker(s) to lay himself out before the fact. There are systems that
have been used to extract values from players in decisions, e.g.,
cognitive mapping [Bonham, Shapiro, and Nozicka, 1976] and others.
However, these must be undertaken and implemented by experts in the
methodology, and definitive results are not yet evident.

The approach used in the cases illustrated here is to make all
the value judgments visible to the decision maker with the key judgments identified. Alternatives can then be evaluated in terms of
the impact that a decision to accept a particular alternative will
have on the judgments. It provides a perspective upon which the
decision-maker(s) can ponder, and it ensures that the judgment cannot be ignored.

Presentation of Information

The second question addresses the manner in which information
can be presented in a meaningful manner and reasonable perspective.
One approach uses benchmarks to provide perspective on similar risks
in society and a second approach is to make visible the entire
gamble upon which the decision impinges.

Benchmarks are not used to imply acceptability, but to provide
references from other, more familiar risks in society. For example,
it is hard to convince a lay person that 80 millirems of exposure
at Three Mile Island is not very harmful, only an increase in
6×10^{-6} per year of getting cancer directly. However, if it is
pointed out that the risk of getting cancer is about the same as
getting lung cancer from smoking four cigarettes in one's lifetime,
a more meaningful perspective and resultant argument can be made.
The benchmarks used in Figure 6 are examples of the same.

The second approach attempts to provide perspective on the
total gamble in which the risks and judgments are imbedded. Question 4 illustrates this concept. However, its practice has not yet
been well applied. Alternate energy systems and oil independence
are only being understood in a total gamble seven years after the
1973 oil embargo and Project Independence. The control of Persian
Gulf oil is suddenly becoming the riskiest situation faced today,
politically, economically, militarily, etc.

The balancing of total gains and losses as opposed to considering only the benefit of reduced risk was addressed previously. This

is one approach to assuring that the total gamble is visualized. However, the implementation of evaluations of this nature is difficult.

CONCLUSION

Specific methodologies for estimating risks are useful for specific situations. Many such methods exist for dealing with ordinary events and objective consequence values. These methods become inadequate when dealing with rare events and value problems. The extension of existing methods to directly deal with these types of problems cannot succeed since the methods and underlying problems they deal with are incompatible.

New approaches, directly addressing the fundamental problems of rare events and value diversity, can be useful, but they must be applied with care and are content dependent. Absolute risk estimates are useful in some cases, i.e., where benchmarks fall outside the uncertainty bands. Relative risk estimates in the comparison of alternatives depend upon the selection of alternatives and the manner in which key issues and value judgments are approached.

As long as the range of methods are content dependent, there will be no general method that can be developed to solve all such problems. Solutions will be more content dependent than method dependent. A methodology for addressing the totality of these problems is but a myth.

REFERENCES

Lowrance, W., Of Acceptable Risk, Kaufman: Los Angeles, 1976.

Okrent, D. and H. Whipple, "An Approach to Societal Risk Acceptable Criteria and Risk Management", UCLA-ENG-7746, June, 1977.

Otway, H., Pahner, D. and Linnerooth, J., "Social Values in Risk Acceptance", IIASA Research Memorandum, 1975, RM-75-54.

Otway, H., "Risk Assessment and Societal Choices", IIASA Research Memorandum, 1975, RM-75-2.

Otway, H. and Fishbein, M., "Public Attitudes and Decision Making", IIASA Research Memorandum, 1977, RM-75-54.

Page, T., "Keeping Score: An Acutarial Approach to Zero-Infinity Dilemmas", Energy Risk Management, (Edited by G. T. Goodman and W. D. Rowe), Academic Press, London, 1979.

U.S. Nuclear Regulatory Commission, "Reactor Safety Study: An
    Assessment of Accident Risks in U.S. Commerical Nuclear
    Power Plants, WAS-1400 (NUREG-75/014), The Commission,
    Washington, D.C., 1975.

"Risk Assessment Service Group Report to the U.S. Nuclear Com-
    mission", H. W. Lewis, Chairman, Ad Hoc Risk Assessment Ser-
    vice Group, September, 1978.

Rowe, W., An Anatomy of Risk, New York, Wiley and Sons, 1977.

Rowe, W. "Assessing Risk to Society", Symposium on Risk Assess-
    ment and Hazard Control, American Chemical Society, New
    Orleans, LA, March 1977.

Rowe, W., "Governmental Regulation of Societal Risk", George
    Washington Law Review, Vol. 45, No. 5, August, 1977.

Slovic, P., Fischhoff, B., S. Lichtenstein, S., Read, S. and
    Combs, B., "How Safe is Safe Enough: A Psychometric Study of
    Attitudes towards Technological Risks and Benefits", Policy
    Science, 1978.

Thedeen, T., "The Problems of Quantification", Energy Risk Manage-
    ment, (Edited by G. T. Goodman and W. D. Rowe) Academic
    Press, London, 1979.

Thomas, K., Mauer, D., Fishbein, M, Otway, H, Hinkle, R. and
    Wimpson, D., "A Comparative Study on Public Beliefs about
    Five Energy Systems", IIASA Research Memorandum, RM-78-XX,
    1978.

Wilson, R., "The FDA Criteria for Assessing Carcinogens Testimony
    at the FDA Hearings on the Sensitivity of Method for Animal
    Food Additives", (Docket No. 77N-0026), 1977.

Wilson, R. and Crouch, E., "Interspecies Comparison of Carcino-
    genic Potency", Journal of Toxicology and Environmental
    Health.

RISK-BENEFIT ANALYSIS IN A MULTIOBJECTIVE FRAMEWORK

Yacov Y. Haimes

Center for Large Scale Systems and Policy Analysis
Case Western Reserve University
Cleveland, Ohio

PREFACE

Over a year ago, the Washington Post commented in its editorial page of October 31, 1979--"Shape up, shape up, shape up!"--on the report issued by the Kemeny Commission:

> A truly unexpected result came out of the Kemeny Commis-
> sion's study of Three Mile Island. A group that set out
> to investigate a technology ended up talking about people.
> The Kemeny report reached two principal conclusions,
> neither of which concerned machine design, communications
> networks, backup systems or any other trappings of nuclear
> technology. In the commission's own words, "It became
> clear that the fundamental problems were people-related
> problems."

What is very clear from the Commission's report--a central theme of this paper--is that the modeling of risk aspects associated with water resources systems, where there is a strong interplay between man and technology, is an intricate task that conventional state-of-the-art modeling methodologies are not adequately equipped to handle. Most importantly, credible modeling of the technological elements of water resources, while essential, will not necessarily lead to a credible overall system model, unless the human system--which affects and is affected by the technological elements--is properly accounted for. The Post's editorial continues:

> Of the many factors that caused the operator errors, the
> commission singled out one: over the years there has been

an almost total preoccupation with equipment and a corresponding failure to appreciate the role of the human
being in the nuclear system. Attention was paid to
"large break" failures of equipment that would happen
very fast and have disastrous consequences. These
accidents could not be affected by reactor operators.
However, the commission found the much greater probability of "small break" equipment failures, which
would happen much more slowly and which "could" be
influenced by operator action, to pose the greater
danger. Because those responsible for nuclear safety
have been hypnotized by equipment, they ignored the
human factor, and Three Mile Island merely illustrated
the consequences.

Specific to the proposed risk analysis methodology is the significance of the role of the decision maker(s) and the associated
people-related problems--a notion well recognized by the Kemeny Commission as well as in this paper.

INTRODUCTION

Most important public policy issues today involve the assessment
of some aspects of risk. The term "risk assessment" is used in this
paper in a broad interpretation given by Rowe [1977a]:

. . . to describe the total process of risk analysis,
which embraces both the determination of levels of risk
and the social evaluation of risks. Risk determination
consists of both identifying risks and estimating the
likelihood of their occurrence. Risk evaluation measures both risk acceptance or the acceptable levels of
societal risk; and risk aversion--methods of avoiding
risks, as alternatives to involuntarily imposed risks.

Public officials and decision makers at all levels of govern-
ment--local, state, regional, and national--are forced to make
public policy decisions without being able to adequately and sufficiently analyze the respective risk impacts and trade-offs associated with their decisions. Thus, the need for respective data is
obvious. It is wise to distinguish, however, between two kinds of
data--information and intelligence. Edward V. Schneier, Jr. [1975],
in testimony before the House Committee on Science and Technology,
offered the following remarks:

Information is, in essence, raw data. It is abundant,
cheap, easy to acquire, sometimes hard to avoid. Witnesses before committees such as this--and I hope I can
avoid their sins--are all too willing to provide it in

great quantities. Intelligence, by which I mean
processed data, data that have been evaluated and
given meaning, is much more difficult to acquire and
much more important to have. Like most scarce commodities, intelligence has value; it confers both
status and power, shapes careers, molds minds.

Information becomes intelligence when it is processed. For a legislator, the problem of processing
is essentially one of investigating facts with their
political significance, of describing who will lose
from a given course of action.

Models, methodologies, and procedures for risk assessment
(referred to, generically, as models in this paper) are aimed at
providing this essential service to decision makers--the processing
of data into intelligence--so that elements of risks associated
with policy decisions may be properly valued, evaluated, and considered in the decision-making process. For such a process to be
viable, several prerequisites should be fulfilled--namely, that

(a) decision-makers be cognizant and appreciative of the im-
    portance of this service; they should also be capable of
    understanding the utility, attributes, and limitations of
    respective models used in risk assessment. Past experience
    does not provide too much encouragement in this respect.

(b) risk assessment models be available, usable, and credible.
    This subject is the central focus of the proposed study.

Rowe [1977b] correctly states that "the process by which risks
are assessed increases in importance as society becomes more cognizant of risk--particularly risk that is inequitably imposed on individuals by the technological activities of their society." Furthermore, Green [1968] makes the point that "the scientific establishment
has developed technology assessment (a decision-making tool weighing
risks against benefits) as a technique for encouraging technological
advance, while protecting the public against new hazards of unpre-
cedented magnitude." These decision-making tools, methodologies,
and models of risk assessment should be evaluated and weighted not
only with respect to their scientific bases but also in terms of
other societal considerations--considerations germane to any sound
public policy decision-making process.

Evaluating the impacts and consequences of public policy decisions involving risk is an imperative step in the process of determining the acceptability of risk. Although this process is known
to be complex, lengthy, and tedious (inasmuch as policy and decision makers must be responsive to a myriad of institutional, legal,
political, historical, and other societal demands and constraints),

the process must be based, to the extent possible, on a firm scientific and technological foundation.

Public policies involving risks are likely to be deemed more acceptable (1) when based on credible scientific and technological information and (2) where sound trade-off and impact analyses have been performed.

Most man-made systems--public and private--are planned, designed, constructed, operated, and modified under conditions of uncertainty and risk. Thus, the assessment, quantification, evaluation, and analysis of risk in the context of systematic policy formation are grounded in the process of decision making. Although several models and methodologies on risk analysis exist, few decision makers understand modeling and systems analysis or appreciate their respective attributes/deficiencies. For example, in classical decision-making literature, such risks are usually reflected in optimization analysis by maximizing (or minimizing) mathematical expectation. This stance assumes that mathematical expectation is, in fact, a suitable scalar measure of the objectives and that all other measures of objectives are incorporated as minimum-level constraints. This assumption further implies that the conditions under which "mathematical expectation" has a valid interpretation are also met. It is a rare risk/decision situation, indeed, where the assumptions necessary for the validity of mathematical expectancy will be met satisfactorily; and it is an even more unusual circumstance when the quantities thus defined adequately represent the risk-minimization objectives of the decision maker [Haimes and Hall, 1977].

Risk analysis for policy- and decision-making purposes is both an art and a science--and yet in emerging phases. Despite a welter of literature on the subject (books and articles are appearing at an exponential rate), numerous fundamental questions remain unanswered:

(a) What is the efficacy of models on risk analysis and
     decision-making?

(b) To what extent are these models and methodologies
     credible?

(c) To what degree does the value of information increase
     with more models on risk?

(d) Who should decide on acceptability of what risks, for
     whom, and on what terms, and why? [W. W. Lowrance,
     1976].

SOURCES OF SKEPTICISM IN MODELING

Generic Modeling Aspects

The difficulties of dealing with the complexity of risk assessment and, particularly, the quantification of risk are familiar to all--policy and decision makers, modelers and analysts, as well as other professionals and the public at large. This complexity is inherent in myriad cases that transcend scientific, technological, economic, political, geographical, legal, and other societal needs, constraints, and limitations. It is not surprising, therefore, that new approaches, models, methodologies, and procedures in risk assessment--which are aimed at assisting the decision makers in these formidable tasks--have filled a real need. On the other hand, policy and decision makers--the ultimate users of these tools--have met these relatively new approaches and risk-assessment methodologies with opinions ranging from outright support to overall skepticism. One may rightfully ask why so many groups have developed not only skepticism but even antagonism against both these analysts and their new analyses [see Haimes, 1976, p. 255]. The following list summarizes some sources of skepticism to modeling and to systems analysis, in general:

- misuse of models and incorrect applications

- insufficient basic science research for credible environmental and social aggregations

- too much model use delegated to people who don't understand models

- insufficient planning and resources for model maintenance and management

- lack of incentives to document models

- over-emphasis on optional use of computers; under-emphasis of efficient use of human resources

- proliferation of models; lack of systematic inventory of available models

- lack of proper calibration, validation, and verification of models

- lack of communication links among modelers, users and affected parties

- models only usable by the developer

- need for models to be recognized as "means," not "ends"

- lack of an interdisciplinary modeling team leading to unrealistic models

- strengths, weaknesses, and limiting assumptions of models often unrecognized by the decision-maker

- insufficient data planning

- lack of consideration of multiple objectives in models

Systems analysis studies (risk assessment studies are no exception) have often been conducted in isolation from the decision makers and commissioned agencies responsible for and charged with implementing any results of these analyses.

In 1976, the General Accounting Office (GAO) extensively studied ways to improve mangement of federally funded computerized models [GAO, 1976]:

> GAO identified 519 federally funded models developed or used in the Pacific Northwest area of the United States. Development of these models cost about $39 million. Fifty-seven of the models were selected for detailed review, each costing over $100,000 to develop. They represent 55 percent of the $39 million of development costs in the models.
>
> Although successfully developed models can be of assistance in the management of Federal programs, GAO found that many model development efforts experienced large cost overruns, prolonged delays in completion, and total user dissatisfaction with the information obtained from the model.

The GAO study classified the problems encountered in model development into three categories: (1) 70% attributable to inadequate management planning, (2) 15% attributable to inadequate management commitment, and (3) 15% attributable to inadequate management coordination. Basically, these problems stem from the simple fact that model credibility and reliability were either lacking or inadequately communicated to management.

During April 28-29, 1977, the National Bureau of Standards (NBS) organized a Workshop on Utility and Use of Large-Scale Mathematical Models [NBS, 1979].

In his welcoming remarks, A. J. Goldman identified major problems facing model developers and users:

. . . Though mainly model developers, we have also
served as methodology contributors, model users, monitors
of modeling efforts, evaluators of models, on occasion
pall-bearers to models--the whole gamut of roles.

This long and sometimes painful history has taught
us a number of disconcerting truths:

- that a model can be conceptually sound, but algorithmically inefficient or inaccurate;

- or algorithmically nifty, but conceptually or
  empirically dubious;

- or technically excellent in every sense, but not useful;

- or, both excellent and useful, but unused;

- or, whether excellent or not, be misused. . . .

These observations have led us to three conclusions--reinforced
by many conversations with colleagues, and strongly corroborated by
recent studies and events:

1. The large federal investment in the development of decision-
   aiding mathematical models has not "paid off," as it can
   and should. Some of the more obvious contributing causes
   involve the absence and/or glut of articulated procedural
   guidelines and professional standards.

2. The attendant disappointments can delay and diminish use
   of the great (perhaps, indispensable) potential of modeling
   to illuminate major public issues and to improve government
   decisions and operations.

3. The professional modeling community should exercise leader-
   ship in identifying and diagnosing underlying problems and
   in moving toward their respective amelioration. Otherwise,
   less palatable prescriptions may be forthcoming from quar-
   ters less sensitive to some of the realities of the modeling
   process--in particular its creative and innovative elements
   and its processual flexibility.

Risk Assessment Models

Risk assessment models, methodologies, and procedures suffer
from most of the same problems identified with modeling in general
and from other related problems. Classic risk-benefit analysis

often:

- overshadows the other considerations--economic, social, aesthe-
  tic, spiritual, etc.

- overemphasizes the "quantifiable" aspects--narrows the decision-
  making process through concentration on easily identifiable
  elements amenable to quantification

- offers only a crutch for tough political decisions

- assumes rational decision-making vs. emotional responses

- assumes political acceptability

- de-emphasizes the importance of "process" in decision making.

Burgeoning public awareness of the dominance of risk and safety
in the determination of contemporary public policy is well reflected
in abundant literature on the subject--both professional journals
and public news media. Public awareness and participation transcend
all aspects of risk and safety--including their measurement, quantification, distribution, and acceptability.

We are aware and appreciate that science and technology play
only one part--albeit a significant part--in the decision-making
process involving risk and safety. William W. Lowrance [1976], in
questioning, who should decide on what risks for whom, and on what
terms, and why? has articulated well the difficulties associated
with risk assessment. The concept of acceptability needs to be given
its proper play among considerations of cost, efficacy, and the
other such components of social decisions. Lowrance states: "We
must hope that the society at large will come to appreciate the cap-
abilities and inherent limitations of science and technology; and
we hope that those in the technical world will come to appreciate
the nonrational nature and overt subtlety of social decisions." In
evaluating the efficacy of risk-assessment models and methodologies,
it is appropriate to keep in mind both Lowrance's advice and that of
Harold P. Green [1975]:

> Scientists and engineers have an important role to
> play in the making of safety determinations. Represen-
> tatives of these disciplines are obviously better
> equipped than others to identify and quantify potential
> risks, and to identify potential benefits. It is ques-
> tionable, however, whether they have any special compe-
> tence to quantify benefits in a manner that can be
> regarded as authoritative in the formulation of public
> policy. No elite group of experts, no matter how broadly

     constituted, has the ability to make an objective and
valid determination with respect to what benefits people
want, and what risks people are willing to assume in
order to have these benefits.

BRIEF LITERATURE REVIEW

    In this section some of the recent literature on risk assessment
is reviewed. My intention is to focus on the applicable methodologies that have been proposed for this purpose. In this context, the
material presented here is based on primary and secondary references.
The primary references directly address the problem of risk assessment, while the secondary references stress the importance of risk
analysis in the decision making process.

    There are several papers and books[*] which provide adequate moti-
vation for a comparative study of risk assessment methodologies. A
central theme through these reports is the classification of risk
mechanisms as endogenous or exogenous--that is, those which have
been humanly engendered or those which are related to natural causes.
Further, a discussion of the various methodologies available for risk
assessment clearly indicates the lack of a unified procedure to select
proper alternatives for risk aversion and reduction. In particular,
there are a myriad of techniques and methodologies which have been
proposed. Thus, the need for an interpretive study to relate these
various techniques is evident.

    In his 1976 report, R. W. Kates [1976] identified four basic
approaches for risk assessment: risk aversion, balanced risk, cost
effectiveness of risk reduction, and risk/benefit balancing. These
four methodologies will briefly be reviewed:

    (a) <u>Risk aversion</u>. These methods are based on the risk-
aversive characteristics of human nature. The main premise of risk
aversion is the avoidance or minimization of a particular risk, with
little or no consideration given to comparison with other benefits
or risks. An example of risk aversion in the United States [Rowe,
1977a] is the establishment of occupational and other safety practices based on identification and evaluation of hazards.

    (b) <u>Balancing risks</u>. The idea here is to compare various
risks and to equalize the consequences which result for each risk

---

*Fischhoff, 1977a; Fischhoff, Slovic, and Lichtenstein, 1979;
Fischhoff, et al., 1978; Kates, 1977; Okrent, 1979; Ritter, 1979;
Slovic, 1978a and 1978b; Vaupel, 1978; Zeckhauser, 1975; Zeckhauser
and Shepard, 1976.

situation. A necessary ingredient in risk balancing is a means for comparing different risk situations. This practice generally requires (i) the definition of an acceptable level of risk and (ii) the determination of risk referents which provide the primary means for comparison.

(c) Cost effectiveness of risk reduction. The basic questions which must be answered are: how much risk is acceptable, and how much are you willing to pay for a particular level of risk aversion? The survey by Rausa [1973] provides some data about the relative worth of various risk aversion strategies.

(d) Benefit-risk analysis. The idea here is to balance the benefits of a particular risk aversion strategy with the actual risks involved--independent of cost factors. This method may be considered a variant of classical benefit-cost analysis, where risk is interpreted as a cost to society. See, for example, the report Perspectives on Benefit/Risk Decision Making [1972] for a more detailed description.

In a more recent paper by Bodily [1980], a method employing multi-attribute decision analysis [as found in the works of Howard, 1968 and 1975; Keeney and Raiffa, 1976; and Raiffa, 1968] is developed for comparison of various risk reducing activities that compete for similar resources. The methodology combines an analysis of the risk aversion characteristics of an individual and the collective analysis of the value of risk reduction to all individuals concerned. The method specifically addresses the following six issues:

• idiosyncratic risk preferences

• nonstandard background risks

• comparison of the value of lifesaving with injury prevention

• "bunching" effects, i.e., the effect of compound risks

• distinction between voluntary and involuntary risks

• possible psychological effects of risk.

In this context, many issues suggested by Kates have been included here.

One of the widely used methods of assessing and communicating the risks of complex systems is a fault tree approach. In this approach all important pathways to failure are listed, followed by a listing of all possible pathways to these pathways, and so on.

Once the desired degree of detail is obtained, the experts assign
probabilities to each of the component pathways, as pointed out by
P. Slovic and B. Fischhoff [1979]. Tree analysis is being used as
a primary methodology in the nuclear reactor safety studies [U.S.
Nuclear Regulatory Commission, 1975]. However, fault tree analysis
has been attacked by critics, who question whether it is valid
enough to be used as a basis for decisions of great consequences
[e.g., Bryan, 1974; Fischhoff, 1977b; Primack, 1975; and Slovic
and Fischhoff, 1979]. Of course, with the occurrence of the Three
Mile Island accident, many skeptical experts now appreciate the concerns voiced by critics of the fault tree method. Slovic and Fischhoff [1979] discuss the various types of error of omission, which,
when committed, may result in unrealistic and incorrect answers by
the fault tree method.

A more elaborate method, which examines benefits as well as
risks, is the revealed performance approach advocated by Starr
[1969]. Although the method of revealed performance is based upon
an intuitively compelling logic, Slovic and Fischhoff [1979] have
discussed some of its drawbacks--among them, that the method:

- assumes that past behavior is a valid predictor of present performance

- it is politically conservative

- makes strong assumptions about the rationality of people's
decision making in the marketplace

- may underweigh risks to which the marketplace responds
sluggishly

- creates difficulty often in developing the measures of risks
and benefits needed for its implementation

As earlier indicated, the classical cost-benefit analysis
method attempts to quantify, in terms of dollars, the expected gains
and losses from a proposed action. The expected cost of a project
is determined (1) by enumerating all aversive consequences that may
arise from its implementation, (2) by assessing the probability that
each will occur, and (3) by estimating the cost or loss to society,
should each such consequence occur. After computing the expected
loss from each possible consequence, the expected loss of the entire
project is computed. An analogous procedure produces an estimate
of expected benefits.

RISK, UNCERTAINTY, AND SENSITIVITY ANALYSIS

The multiobjective aspects of risk assessment have been drama-
tized by two presidential directives.

I. On July 12, 1978, President Carter directed the U.S. Water
    Resources Council to carry out a thorough evaluation of cur-
    rent agency practices for making benefit and cost calculations
    including uncertainty and risk of costs and benefits associ-
    ated with Federal water resources planning.

    On December 14, 1979, the U.S. Water Resources Council pub-
    lished its findings in the Federal Register. Section 713.31,
    "Risk and Uncertainty--Sensitivity Analysis" provides "guid-
    ance for the evaluation of risk and uncertainty in the formu-
    lation of water resources management and development plans."

II. On February 17, 1981, President Reagan issued Executive Order
    12291 concerning Federal Regulation. Section 3, "Regulatory
    Impact Analysis and Review," requires that the following
    information must be contained in the Regulatory Impact Analy-
    sis:

    (1) A description of the potential benefits of this rule,
         including any beneficial effects that cannot be quanti-
         fied in monetary terms, and the identification of those
         likely to receive the benefits;

    (2) A description of the potential costs of the rule, includ-
         ing any adverse effects that cannot be quantified in
         monetary terms, and the identification of those likely
         to bear the costs;

    (3) A determination of the potential net benefits of the
         rule, including an evaluation of effects that cannot be
         quantified in monetary terms;

    (4) A description of alternative approaches that could sub-
         stantially achieve the same regulatory goal at lower
         cost, together with an analysis of this potential bene-
         fit and costs and a brief explanation of the legal rea-
         sons why such alternatives, if proposed, could not be
         adopted; and

    (5) Unless covered by the description required under para-
         graph (4) of this subsection, an explanation of any legal
         reasons why the rule cannot be based on the requirements
         set forth in Section 2 of this Order.

Central to the above directives is the identification, quantification, and inclusion in the analysis of all noncommensurable costs and benefits. Both presidential directives emphasize the importance of addressing consequences which cannot be quantified in monetary terms--backbone of the proposed multiobjective framework.

The revised Water Resources Council's Principles, Standards and Procedures (P, S & P) requires the evaluation of risk and uncertainty in the formulation of water resources management and development plans [U.S. Water Resources Council, 1980]. For the purpose of this presentation, it is convenient to use the Council's definitions of risk and uncertainty:

(a) Risk. Situations of risk are conventionally defined as those in which the potential outcomes can be described in reasonably well-known probability distributions. For example, if it is known that a river will flood to a specific level on the average of once in 20 years, a situation of risk, rather than uncertainty, exists.

(b) Uncertainty. In situations of uncertainty, potential outcomes cannot be described in objectively known probability distributions.

In addition, the U.S. Water Resources Council's above document [1980] identifies two major sources of risk and uncertainty:

1. Risk and uncertainty arise from measurement errors and from the underlying variability of complex natural, social, and economic situations. If the analyst is uncertain because the data are imperfect or the analytical tools crude, the plan is subject to measurement errors. Improved data and refined analytic techniques will obviously help minimize measurement errors.

2. Some future demographic, economic, hydrologic, and meteorological events are essentially unpredictable because they are subject to random influences. The question for the analyst is whether the randomness can be described by some probability distribution. If there is a historical data base that is applicable to the future, distributions can be described or approximated by objective techniques.

If there is no such historical data base, the probability distribution of random future events can be described subjectively, based upon the best available insight and judgment.

The ultimate efficacy of risk assessment in water resources planning and management is the assistance it provides planners and respective decision makers in:

(a) identifying the sources of risk and uncertainty associated with exogenous variables and events such as demographic, economic, hydrologic, meteorological, environmental, institutional, and political factors.

(b) quantifying the input-output relationships with respect to the randomness of these exogenous variables and events to the degree possible and feasible, given the constraints on data and information.

(c) quantifying, to the degree possible and feasible, the potential or probable impacts of risk and uncertainty and their associated trade-offs on alternative policy decisions.

(d) facilitating a decision-making process where decision makers can make the utmost scientific use of intelligence about risk and uncertainty related to trade-off and decision analysis of human factors.

It is instructive to articulate, at this stage, the difference between process and methodologies in risk assessment.

The process of risk assessment is the aggregation of interactions with decision makers in the application of risk assessment approaches. (These interactions involve trade-off analysis and the exercise of value judgments.)

The methodologies of risk assessment are the techniques utilized in the scientific approach of estimating probabilities and performing risk assessment (excluding the application of value judgments)--an integral part of the process.

It is also noteworthy that the risk assessment process--the setting of value judgments and parameters--is critically important, because it facilitates the educational process of the analysts and decision makers and their understanding of the methodologies. In turn, the methodologies serve as important stimuli for discussion (in addition to their contribution to quantification and the transfer of information into intelligence), even if the methodologies themselves are not very good. Clearly, methodologies are a necessary condition for a credible and viable risk assessment process, but are, by no means, sufficient.

This process can help to identify and articulate the issues upon which there is an agreement among decision makers, and also those for which there is no agreement. The process also helps to make the implicit explicit. This outcome, however, may embarrass decision makers under certain circumstances.

Most methodologies in risk-benefit analysis address <u>risk</u> aspects--events in which potential outcomes can be described in reasonably well-known probabilities. In contrast, the remainder of this paper deals with the <u>uncertainty</u> aspects in risk-benefit analysis--events in which the potential outcomes cannot be described in objectively known probabilities.

Usually, it has been common to fit all benefits, costs, and risks into a simple criterion. In contrast, the methodology presented in this paper preserves the noncommensurable attributes of benefits, costs, and risks in the quantitative evaluation and trade-off analysis. This is accomplished via the use of the Surrogate Worth Trade-off (SWT) method [Haimes and Hall, 1974]--a method for analyzing and optimizing multiple noncommensurable objective functions.

At this stage, it is instructive to briefly review the SWT method and its attributes.

The Surrogate Worth Trade-off (SWT) Method

Decision making is an integral part of our daily lives, and it ranges in complexity from simple to very complex situations involving multiple objectives. Because of the diversity of situations in which multiobjective decision problems can arise and because of the multiplicity of factors that are involved, the bulk of literature on the subject which has been produced since early 1960 is large in number and diverse in emphasis and style of treatment. The general indication is that this trend will continue. Theoretical and methodological developments are based on different viewpoints that reflect the diversity of disciplines involved, and focus on different aspects of multiobjective decision problems.

To describe a multiobjective decision problem, one needs to clearly specify:

• the decision-making unit, one of whose components is the decision maker

• a set of objectives and its hierarchy

• an appropriate set of attributes and complete objective-attribute relationships if they are not obvious

• the decision situation

• the decision rule

The Surrogate Worth Trade-off (SWT) method--a methodology for analyzing and optimizing multiple noncommensurable objective functions with single or multiple decision-makers--has been extensively discussed in the literature.* The purpose of this section is to briefly summarize the basic steps and attributes of the SWT method, and, thus, to add completeness to this presentation.

Fundamental to multiobjective analysis is the Pareto optimal concept, which is also known as a non-inferior solution. Qualitatively, a non-inferior solution of a multiobjective problem is one in which any improvement of one objective function can be achieved only at the expense of degrading another.

To define a non-inferior solution mathematically, consider the following multiobjective optimization problem which is also known as a vector optimization problem:

$$\underset{x \in X}{\text{minimize}} \ \{f_1(\underline{x}), f_2(\underline{x}), \ldots, f_n(\underline{x})\} \tag{1}$$

where

$$X = \{\underline{x} | \underline{x} \in R^N, \ g_i(\underline{x}) \leq 0 \quad i = 1, \ldots, m\}$$

and $f_i(\underline{x})$ and $g_i(\underline{x})$ are properly defined objective functions and constraints, respectively.

Definition: A decision x^* is said to be a non-inferior solution to the problem posed by the systems (1), if and only if there does not exist another $\bar{x}$ so that $f_j(\bar{x}) \leq f_j(x^*)$, $j = 1, 2, \ldots, n$, with strict inequality holding for at least one j.

Essentially the SWT method consists of the following three steps.

Step 1. Generate a representative subset of noninferior solutions using the ε-constraint method with f_k as the primary objective. I.e., we solve

$$\underset{x \in X}{\text{min}} \ f_k(\underline{x}) \tag{2}$$

subject to

---

*Haimes and Hall, 1974; Hall and Haimes, 1976; Haimes, 1977; Haimes and Chankong, 1979; Haimes et al., 1980; and Tarvainen and Haimes, 1980.

$$f_j(\underline{x}) \leq \varepsilon_j \quad \text{for all} \quad j = 1,\ldots,k-1,k+1,\ldots,n \qquad (3)$$

for as many values of ε_j (which may be varied from $f_j^* = $ min $f_j(\underline{x})$ upward) as desired.
$\underline{x} \in X$

It has been shown [Haimes, Lasdon, and Wismer, 1971] that the two problems posed by (1) and (2)-(3) are equivalent. In addition, both Chankong [1977] and Haimes and Chankong [1979] illustrate that, under appropriate assumptions, solutions of (2)-(3) are, indeed, noninferior.

Step 2. For each noninferior solution, obtain trade-off information between f_k and each of the other criteria. In the process of solving (2)-(3) either analytically or by existing optimization algorithms, this "trade-off" information can be easily obtained by merely observing that the optimal Kuhn-Tucker multipliers correspond to the binding constraints of the form (3). Under appropriate conditions, these multipliers will represent "local" trade-off between f_k and the corresponding f_j. For example, if, at the optimal solution $\underline{x}^0$ of (2)-(3), constraints (3) are binding for all $j \neq k$, and $\lambda_{kj} > 0$ for all $j \neq k$ are the corresponding optimal Kuhn-Tucker multipliers, then it can be shown that

$$\lambda_{kj} = - \frac{\partial f_k(\underline{x}^0)}{\partial f_j(\underline{x}^0)} \qquad (4)$$

when all other f_i, $i \neq j$, and $i \neq k$ are held constant at $f_i(\underline{x}^0)$. An extension of the theoretical basis of this trade-off interpretation of optimal Kuhn-Tucker multipliers can be found in Chankong [1977] and Haimes and Chankong [1979].

Step 3. The decision maker, DM, is supplied with trade-off information obtained from Step 2 as well as the levels of all criteria. The DM then expresses his ordinal preference about whether or not (and by how much) he would like to make such a trade at the level. The surrogate worth function is then constructed from this information.

The SWT method has been extended to problems with multiple decision makers [Haimes and Hall, 1977], and successfully applied to river basin planning [Haimes, Das, and Sung, 1979].

Let, for example, $\underline{x}^0$ be a noninferior solution generated in
Step 1, and $\lambda_{kj}(\underline{x}^0)$ be a trade-off between f_k and f_j for each
$j = 1,\ldots,k-1,k+1,\ldots,n$ (as determined in Step 2), when we move
from $\underline{x}^0$ to some other noninferior solution close to $\underline{x}^0$. Knowing
all $f_j(\underline{x}^0)$ and $\lambda_{kj}(\underline{x}^0)$, the DM may express his degree of preference in trading $\lambda_{kj}(\underline{x}^0)$ units of f_k with one unit of f_j, by
assigning an integer value between +10 and -10 to W_{kj}, which is
called the "surrogate worth." High positive values of W_{kj}, say
"+10," means the DM has great desire to make that trade and vice
versa. $W_{kj} = 0$ indicates "indifference" on the part of the DM
toward making that particular trade. After W_{kj} is estimated for
each $j \neq k$ and for several points in the representative subset
of noninferior solutions, the surrogate worth function $W_{kj}(\cdot)$
[which relates W_{kj} to f_j for all $j \neq k$ and λ_{kj}] is constructed for each $j \neq k$. Then an indifference point $\underline{\varepsilon}^* =$
$(f_1^*,\ldots,f_{k-1}^*,f_{k+1}^*,\ldots,f_n^*)$ at which $W_{kj} = 0$ for all $j \neq k$
could be found, so that the most-preferred solution $\underline{x}^*$ could be
reconstructed by solving the ε-constraint problem with f_k as the
primary objective and $\underline{\varepsilon}^*$ as the constraints [Everett, 1963].

A basic goal of the Surrogate Worth Trade-off (SWT) method is
to develop a set of Pareto optimal solutions and consistent trade-
off functions between the competing objectives, then arrive at the
decision-makers' preferences. An important characteristic of the
SWT method is that it properly leaves to the specialized analysts
the quantitative-predictive (scientific) aspects of evaluation, but
clearly gives the decision maker(s) the right and responsibility
for evaluating the merits of improving any one objective at the
expense of any other, given the associated quantitative levels of
achievement of all objectives. Because of this characteristic, the
SWT method lends itself well either to simulating the likely outcome
of the multiple-member decision process (given the characteristics
of each decision maker) or to assisting the multiple-member decision
group in identifying and focusing efficiently on the issues implicit
in the problem's structure and public constituency.

Several extensions of the SWT method [Haimes and Hall, 1974]
have been reported in the literature, and include:

(a)   multiobjective optimization of dynamic systems--optimal
      control problems [Haimes, Hall, and Freedman, 1975]

(b)   multiple decision-makers [Hall and Haimes, 1976; and
      Haimes, 1977]

(c)   the incorporation of risk assessment in a multiobjective
      framework [Haimes, Hall, and Freedman, 1975; and Haimes
      and Hall, 1977]

(d)   total and partial trade-offs, and necessary and sufficient
      conditions for noninferiority [Chankong, 1977; and Haimes
      and Chankong, 1979]

(e)   the Interactive Surrogate Worth Trade-off (ISWT) method
      [Chankong, 1977; and Chankong and Haimes, 1978]

(f)   the Multiobjective Statistical Method (MSM)--incorporat-
      ing probabilistic elements into the SWT method [Haimes,
      1980]

(g)   the SWT method as part of a higher-level coordinator for
      hierarchical-multiobjective models [Tarvainen and Haimes,
      1980; and Haimes and Tarvainen, 1980]

In addition to the above extensions, numerous applications of
the SWT method to case studies are reported in literature throughout
the U.S. and abroad.

Sensitivity as an Index for Risk

Sensitivity analysis has long been considered an integral part
of all water resources studies, in recognition of the inherent randomness of hydrologic and socio-economic events. In the Revised
Principles and Standards for Water and Related Land Resources
Planning, Level C [Fed. Register, 1980], the U.S. Water Resources
Council states: "The planner's primary role in dealing with risk
and uncertainty is to identify the areas of sensitivity and describe
them clearly so that decisions can be made with knowledge of the
degree of reliability of available information." The ideas and
methodology advocated in this section are both in congruence with
and in support of the above statement. While there is near unanim-
ity among water planners regarding the imperativeness of sensitivity
analysis, the ways and means of conducting sensitivity analysis and
integrating it into the overall study or plan are still debatable.
In particular, the trend has been to use sensitivity analysis as a
post-study and extrinsic evaluation (of the study), rather than as
a genuine component of the study in terms of trade-off analysis of

the risks, costs, and benefits--as is proposed here [Haimes, Hall, and Freedman, 1975; and Haimes and Hall, 1977].

Inherent in every water resources policy, plan, or model is the existence of a number of dominant exogenous variables with randomness of either known or unknown probabilities. When the respective probability distribution functions are not known, a sensitivity function/index of the model output(s) with respect to one or more of these exogenous variables can serve as a surrogate risk function. Then, treating these risk functions as objective functions along with the original model's objective function(s) yields a multiobjective optimization problem, amenable to solution methodologies such as the SWT method discussed in the previous section.

Model Formulation. Consider the capacity expansion problem of a water supply system confronting a respective planning agency. One objective might be to minimize the cost of the additional water supply system, while providing an acceptable level of water supply and satisfying socio-economic and other constraints. The optimal capacity expansion of water resources systems has been widely cited in the literature [Butcher, Haimes, and Hall, 1969; Haimes and Nainis, 1974; Nainis and Haimes, 1975; and Haimes, 1977]. The worth of streamflow data in water resources planning, with respect to the capacity expansion problem, was also discussed in recent literature using a multiobjective optimization framework [Haimes, Craig, and Subrahmanian]. The risk and uncertainty aspects of water supply systems are addressed herein.

Let

$y(\underline{x},\underline{\alpha})$ denote the additional water supply provided through the capacity expansion of the existing water system

where,

$\underline{x}$ denotes the vector of decision variables and

$\underline{\alpha}$ denotes the vector of exogenous variables.

Furthermore, let

$f_1(\underline{x},y,\underline{\alpha})$ denote the overall cost function of the project.

Since $y = y(\underline{x},\underline{\alpha})$, the cost function can be simplified:

$$f_1(\underline{x},\underline{\alpha}) = f_1(\underline{x},\underline{y},\underline{\alpha}) .$$

Thus, the overall optimization problem can be written as:

$$\text{minimize} \quad f_1(\underline{x},\hat{\underline{\alpha}}) \qquad\qquad\qquad (5)$$
$$\underline{x}\in X$$

where

$$X = \{\underline{x}|\underline{x} \in R^N, g_i(\underline{x},\hat{\underline{\alpha}}) \le 0, \quad i = 1,2,\ldots,m\} \qquad (6)$$

and $f_1(\underline{x},\underline{\alpha})$ and $g_i(\underline{x},\underline{\alpha})$ are properly defined objective functions and constraints, respectively. The notation $\hat{\underline{\alpha}}$ connotes the nominal value of $\underline{\alpha}$.

For pedagogical purposes, assume that α_1, the first element of the vector $\underline{\alpha}$, is a dominantly influential exogenous variable, e.g., water demand for a rapidly developing region. It is also assumed that the α_1 cannot be described in terms of an objectively known probability distribution function. For simplicity, also assume that the respective surrogate risk function can be derived on the basis of the sensitivity of the overall development cost of the water project, $f(\underline{x},\underline{\alpha})$, to variations in α_1. This reasonable assertion is based on the fact that the cost function $f_1(\cdot)$ is a monotone, nondecreasing function of the planned capacity of the water project, $y(\cdot)$. A primitive sensitivity function [Rarig, 1976; and Rarig and Haimes, 1981] can be derived as

$$\frac{\partial f_1(\underline{x},\underline{\alpha})}{\partial \alpha_1} .$$

The respective surrogate risk function, $f_2(\underline{x},\underline{\alpha})$ is derived in (7):

$$f_2(\underline{x},\underline{\alpha}) = \left(\frac{\partial f_1(\underline{x},\underline{\alpha})}{\partial \alpha_1}\right)^2 \qquad\qquad (7)$$

The overall multiobjective optimization problem, where uncertainty is intrinsically considered in the decision-making process, is given by (8):

$$\text{minimize} \quad \begin{pmatrix} f_1(\underline{x},\hat{\underline{\alpha}}) \\ \\ f_2(\underline{x},\hat{\underline{\alpha}}) \end{pmatrix} \qquad\qquad (8)$$
$$\underline{x}\in X$$

<u>Example Problem</u>. Consider the following model:

$$\text{minimize} \ \{f_1(\underline{x},\alpha) = 2(x_1-1)^2 - 2\alpha(x_1+x_2) + (x_2-2)^2 - \alpha^2\} \quad (9)$$
$$\underline{x}$$

where for simplicity, α is assumed to be a scalar, and all the system's constraints other than $x_1 \geq 0$, $x_2 \geq 0$ are ignored.

The surrogate uncertainty function $f_2(\underline{x},\alpha)$ is given in (10):

$$f_2(\underline{x},\alpha) = \left(\frac{df_1(\underline{x},\alpha)}{d\alpha}\right)^2 = \left[-2(x_1+x_2) - 2\alpha\right]^2 \quad (10)$$

The resulting multiobjective optimization problem is then:

$$\text{minimize} \ \left\{ \begin{array}{l} 2(x_1-1)^2 - 2\alpha(x_1+x_2) + (x_2-2)^2 - \alpha^2 \\[2mm] 4(x_1+x_2)^2 + 8\alpha(x_1+x_2) + 4\alpha^2 \end{array} \right\}_{\alpha=\hat{\alpha}} \quad (11)$$
$$\underline{x}$$

For computational purposes, assume $\hat{\alpha} = 4$.

Transfer (11) into the ε-constraint formulation:

$$\left. \begin{array}{l} \text{minimize} \ f_1(\underline{x},\alpha) \\ \underline{x} \\[4mm] \text{subject to} \\[4mm] f_2(\underline{x},\alpha) \leq \varepsilon_2 \end{array} \right\} \quad (12)$$

Form the Lagrangian $L(\cdot)$ for (12):

$$L(\underline{x},\alpha,\lambda_{12}) = f_1(\underline{x},\alpha) + \lambda_{12}[f_2(\underline{x},\alpha) - \varepsilon_2] \quad (13)$$

substituting (11) into (13) yields:

$$L(\underline{x},\alpha,\lambda_{12}) = 2(x_1-1)^2 - 2\alpha(x_1+x_2) + (x_2-2)^2 - \alpha^2$$
$$+ \lambda_{12}[4(x_1+x_2)^2 + 8\alpha(x_1+x_2) + 4\alpha^2 - \varepsilon_2] \quad (14)$$

Necessary conditions for stationary points are:

$$\frac{\partial L(\cdot)}{\partial x_1} \geq 0; \quad x_1 \frac{\partial L(\cdot)}{\partial x_1} = 0; \quad x_1 \geq 0 \tag{15}$$

and

$$\frac{\partial L(\cdot)}{\partial x_2} \geq 0; \quad x_2 \frac{\partial L(\cdot)}{\partial x_2} = 0; \quad x_2 \geq 0 \tag{16}$$

$$\frac{\partial L(\cdot)}{\partial \lambda_{12}} \leq 0, \quad \lambda_{12} \frac{\partial L(\cdot)}{\partial \lambda_{12}} = 0, \quad \lambda_{12} \geq 0$$

Assuming that some of the Pareto optimal solutions are not on the boundary at $x_1 = 0$, $x_2 = 0$ yields:

$$\frac{\partial L(\cdot)}{\partial x_1} = 4(x_1-1) - 2\alpha + 8\lambda_{12}(x_1 + x_2) + 8\lambda_{12}\alpha = 0$$

or

$$\lambda_{12} = \frac{2 - 2x_1 + \alpha}{4(x_1 + x_2 + \alpha)} \tag{17}$$

Note that for $\hat{\alpha} = 4$, $\lambda_{12} > 0$ only if $x_1 < 3$.

$$\frac{\partial L(\cdot)}{\partial x_2} = -2\alpha + 2(x_2-2) + 8\lambda_{12}(x_1 + x_2) + 8\lambda_{12}\alpha = 0$$

or

$$\lambda_{12} = \frac{2 - x_2 + \alpha}{4(x_1 + x_2 + \alpha)} \tag{18}$$

Note that for $\hat{\alpha} = 4$, $\lambda_{12} > 0$ only if $x_2 < 6$.

Equalizing (17) and (18) yields

$$\left. \begin{array}{l} x_2 = 2x_1 \\[4pt] \text{for } 0 \leq x_1 \leq 3; \quad 0 \leq x_2 \leq 6 \end{array} \right\} \tag{19}$$

All above pairs, $x_2 = 2x_1$, corresponding to $\lambda_{12} > 0$ yield Pareto optimal solutions (except for $x_1 = 3$ and $x_2 = 6$). These solutions are given in Table 1 for $\hat{\alpha} = 4$ and depicted in the functional space in Figure 1.

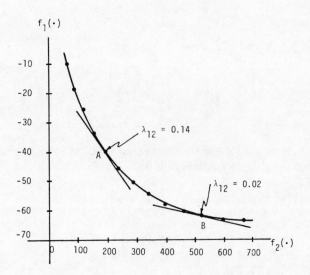

Fig. 1. Pareto optimal policies

Table 1

x_1	x_2	f_1	f_2	λ_{12}
0	0	-10	64	.38
.5	1.0	-26.5	121	.23
1.0	2.0	-40	196	.14
1.5	3.0	-50.5	289	.09
2.0	4.0	-58	400	.05
2.5	5.0	-62.5	529	.02
2.99	5.98	-63.99	673	0.0004
3.0	6.0	-64	676	0

Policy Analysis. The above quantitative analysis should now
be incorporated with the subjective and judgmental decisions of the
planner(s)/decision maker(s). The trade-offs to be made are between
two extreme options:

(a) Business-as-Usual Option--Develop the water resources system
     without regard for the potential impacts of variations on the
     projected demand α from its assumed nominal value, $\hat{\alpha} = 4$.
     This option translates into a Pareto optimal decision
     $\underline{x}^* = (3,6)$. The corresponding values of the objective func-
     tions and the trade-off to $\underline{x}^*$ are:[†]

$$\left.\begin{array}{l} f_1(\underline{x}^*;\hat{\alpha}) = -64 \\[2mm] f_2(\underline{x}^*;\hat{\alpha}) = 676 \\[2mm] \lambda_{12}(f_2(\underline{x}^*,\hat{\alpha})) = 0 \end{array}\right\} \tag{20}$$

     In summary, this option essentially ignores the sensitiv-
ity objective function, $f_2(\underline{x};\hat{\alpha})$. Where

$$\underset{\underline{x}\in X}{\text{minimize}} \quad f_1(\underline{x},\hat{\alpha}) = f_1(\underline{x}^*,\hat{\alpha}) \tag{21}$$

(b) Conservative/Risk Aversion Option--Develop the water resources
     system without regard to its cost. This option translates into
     a Pareto optimal decision $\hat{\underline{x}} = (0,0)$. The corresponding values
     of the objective functions and the respective trade-off to $\hat{\underline{x}}$
     are:

$$\left.\begin{array}{l} f_1(\hat{\underline{x}};\hat{\alpha}) = -10 \\[2mm] f_2(\hat{\underline{x}},\hat{\alpha}) = 64 \\[2mm] \lambda_{12}(f_2(\hat{\underline{x}};\hat{\alpha})) = 0.38 \end{array}\right\} \tag{22}$$

     In summary, this option essentially ignores the cost ob-
jective function, $f_1(\underline{x},\hat{\alpha})$, where

$$\underset{\underline{x}\in X}{\text{minimize}} \quad f_2(\hat{\underline{x}};\hat{\alpha}) = f_2(\hat{\underline{x}};\hat{\alpha}) \tag{23}$$

     Clearly, a realistic and viable risk-benefit analysis (reducing
risk of water shortages due to uncertainties in the projected demand

---

[†]For convenience, the Pareto optimal boundary points (2.999, 5.999)
are rounded to (3.6).

function and, at the same time, reducing the overall cost of developing the water supply system) will consider the trade-offs between the two extreme options, with the likelihood of adopting neither extreme. Namely, the decision, $\underline{x}$, will be somewhere between $\underline{x}^*$ and $\underline{\hat{x}}$.

$$\underline{\hat{x}} \leq \underline{x} \leq \underline{x}^* \tag{24}$$

To determine the specific policy associated with acceptable risk, referred to in the section on the SWT method as preferred Pareto optimal solution, the second phase of the SWT method is used--interaction with the decision maker(s) via the surrogate worth function.

Acceptable Risk. The attempt to determine the acceptability of risk involving public policy--as is the case in water resources systems--is undoubtedly the most exacting and vulnerable task of the entire process of risk-benefit analysis. In earlier sections, it was argued that the fundamental problem of acceptability of risk may never be answered satisfactorily (with the exception of limited case-by-case studies). Nevertheless, the planner(s)/decision-maker(s) in the above example must either explicitly or implicitly determine the level of project capacity expansion, and, hence, the associated and respective level of risk of prospective water shortages. The SWT method, by quantitatively transferring available information into intelligence within a multiobjective framework-- generating Pareto optimal solutions and their respective trade-offs, and subsequently integrating with this intelligence, via the surrogate worth functions, the subjective and judgmental preferences of the decision maker(s)--facilitates the process of determining the acceptability of risk. In this process, the planner(s)/decision-maker(s) base their decisions not only on the absolute levels of f_1 vs. f_2 (project cost/level of supply versus risk of water shortages), but also on the respective trade-offs of these levels.

Consider, for example, points (policies) A and B in Figure 1. Policy A translates into a high cost and a low level of risk with a respective large trade-off ($f_1 = -40$, $f_2 = 196$ and $\lambda_{12} = 0.14$, see Table 1). This means that, at this point, an increase of risk can be achieved at a relatively large cost reduction. Policy B, on the other hand, translates into a low cost and a high level of risk with a respective low trade-off ($f_1 = 62.5$, $f_2 = 529$, and $\lambda_{12} = 0.02$). This means that, at this point, a further reduction of risk can be achieved by a relatively low additional cost. Through the surrogate worth function, which is imbedded in the SWT method, the planner(s)/decision-maker(s) can articulate their preferences and reach a preferred Pareto optimal solution, also known as the best compromise solution.

Further Analysis. It is instructive to further analyze the interpretations of $\underline{x}^*$ and $\hat{\underline{x}}$ in the decision-making context. Note that the level of capacity expansion, $y(\cdot)$, and its associated cost function, $f_1(\cdot)$, are explicit functions of the projected demand, α. The above risk-benefit analysis and the generation of Pareto optimal solutions were made on the basis of the sensitivity of the outcome in the neighborhood of the nominal value of the demand, $\hat{\underline{x}} = \hat{\underline{x}}(\hat{\alpha})$. Investigating the behavior of the cost function $f_1(\underline{x},\alpha)$ [and thus the $y(\underline{x},\alpha)$] for the two distinct policies $\underline{x}^*$ and $\hat{\underline{x}}$ as a function of water demand yields the following results:

Substituting $\underline{x}^*$ and $\hat{\underline{x}}$ into equation (9),

$$f_1(\underline{x},\alpha) = 2(x_1-1)^2 - 2\alpha(x_1 + x_2) + (x_2-2)^2 - \alpha^2$$

yields:

$$f_1(\underline{x}^*,\alpha) = 2(x_1^* - 1)^2 - 2\alpha(x_1^* + x_2^*) + (x_2^* - 2)^2 - \alpha^2$$
$$f_1(\underline{x}^*,\alpha) = 24 - 18\alpha - \alpha^2 \qquad (25)$$

$$f_1(\underline{x},\alpha) = 2(\hat{x}_1-1)^2 - 2\alpha(\hat{x}_1 + \hat{x}_2) + (\hat{x}_2 - 2)^2 - \alpha^2$$
$$f_1(\hat{\underline{x}},\alpha) = 6 - \alpha^2 \qquad (26)$$

Equations (25) and (26) represent the behavior of the cost function [and the respective capacity expansion level $y(\cdot)$] following the business-as-usual option, $\underline{x}^*$, or the conservative/risk aversion option, $\hat{\underline{x}}$, respectively, as functions of the projected demand α. These equations are evaluated for various values of α, as presented by Table 2 and depicted in Figure 2. Note that, at $\alpha = 4$, the slope of

$$f_1(\hat{\underline{x}},\alpha), \quad \left.\frac{df_1(\hat{\underline{x}},\alpha)}{d\alpha}\right|_{\alpha=\hat{\alpha}}$$

is -8[†]. Namely, changes in α in the neighborhood of $\hat{\alpha}$ have minimal effect on $f_1(\cdot)$ or $y(\cdot)$. Stated differently, if the project were planned--assuming a nominal level of demand, $\alpha = \hat{\alpha} = 4$, and following the conservative/risk aversion policy $\hat{\underline{x}}$ --the risk of having water shortages due to uncertainties in the demand projection is minimal.

[†]This value is not zero because of the constraints $x_1 \geq 0$, $x_2 \geq 0$.

Table 2

α	$f_1(x^*,\alpha)$	$f_1(\hat{x},\alpha)$
-20	- 16	
-18	24	
-16	56	
-14	80	
-12	96	-207
-10	104	-147
- 8	104	- 95
- 6	96	- 51
- 4	80	- 15
- 2	56	13
0	24	33
2	- 16	45
4	- 64	49
6	-120	45
8	-184	33
10		13
12		- 15
14		- 51
16		- 95
18		-147
20		-207

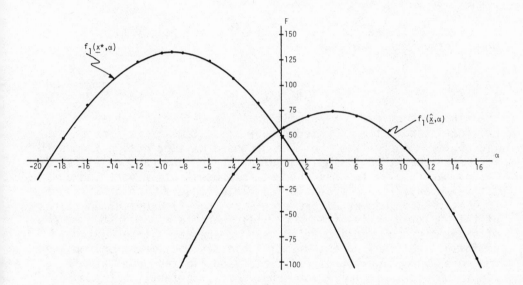

Fig. 2. $f_1(\hat{\underline{x}},\alpha)$ and $f_1(\underline{x}^*,\alpha)$ vs. α

On the other hand, at $\alpha = 4$, the slope of $f_1(\underline{x}^*, \alpha)$ is large:

$$\frac{df_1(\underline{x}^*, \alpha)}{d\alpha}\bigg|_{\alpha=\hat{\alpha}} = -26.$$

Namely, changes in α in the neighborhood of $\hat{\alpha}$ might have drastic impacts on $f_1(\cdot)$ and $y(\cdot)$. Stated differently, if the project were planned--assuming a nominal level of demand, $\alpha = \hat{\alpha} = 4$, and following the business-as-usual policy, $\underline{x}^*$ --there is a very high risk of water shortages attributable to uncertainties in demand projection.

An ultimate responsible (and responsive) policy might be somewhere between the two externals-- $\hat{\underline{x}} \leq \underline{x} \leq \underline{x}^*$.

EPILOGUE

The subject of risk-benefit analysis will continue to be a critical agenda item in water resources systems as well as in public policy. The efficacy of the proposed multiobjective framework for risk-benefit analysis is grounded in the dominant role that decision makers play in this process--a process of integrating quantitative scientific and technological intelligence with qualitative subjective and judgmental preferences. The efficacy and usefulness of this multiobjective framework will increase as (a) the sources of skepticism in modeling (discussed in the Literature Review) diminish, (b) the concept of multiobjective optimization replaces the more conventional, yet restricted and limited, concept of simple objective optimization, and, most importantly, (c) more analysts recognize that a viable risk-benefit decision-making process can only be attained within a proper framework which enhances dialogue and interactions among analysts, decision-makers, and the affected constituencies.

ACKNOWLEDGEMENTS

The concepts and the methodology presented in this paper have been developed over the years primarily through our research on multiobjective optimization. Therefore, special thanks are due to Dr. W. A. Hall who has been collaborating and contributing with the author for the last sixteen years. The author also thanks Drs. K. Loparo and S. Sorooshian for their contributions to the section on Risk, Uncertainty, and Sensitivity Analysis; Dr. M. A. H. Ruffner for her editorial contributions; and Dr. V. Covello for his advice and encouragement. Thanks are also due to D. Gemperline,

V. Chankong, and P. Zwick for their assistance in the numerical example. Research for this paper was supported in part by Environmental Systems Management, Inc.

REFERENCES

Bodily, Samuel E., 1980, Analysis of risks to life and limb, Operations Res., 1:156-175.

Bryan, W. B., Feb. 1, 1974, testimony before the Subcommittee on State Energy Policy, Committee on Planning, Land Use, and Energy, California State Assembly.

Butcher, W. S., Haimes, Y. Y., and Hall, W. A., 1969, Dynamic programming for the optimal sequencing of water supply projects, Water Resources Res., 5(6):1196-1204.

Chankong, V., 1977, Multiobjective Decision Making Analysis, Ph.D. Dissertation, Case Western Reserve University, Cleveland, Ohio.

Chankong, V., and Haimes, Y. Y., 1978, The interactive Surrogate Worth Trade-off (ISWT) method for multiobjective decision-making, in: "Multicriteria Problem Solving," Stanley Zoints, ed., Springer-Verlag, New York.

Committee on Science and Technology, U.S. Congress, Risk/benefit analysis in the legislative process, Joint Hearings, 96th Congress, July 24-25, 1979, Committee Report No. 71, published March 1980.

Everett, H., III, 1963, Generalized Lagrange multiplier method for solving problems of optimum allocation of resources, Operations Res., 11:399-418.

Federal Register, September 23, 1980, 45(190):64391.

Fischhoff, Baruch, 1977a, Cost benefit analysis and the art of motorcycle maintenance, Policy Sci., 8:177-202.

Fischhoff, B., 1977b, Perceived informativeness of facts, J. Experimental Psych., Human Perception and Performance, 3:349-358.

Fischhoff, Baruch, Hohenemsen, Christoph, Kosperson, Roger E., and Kates, Robert W., 1978, Handling hazards, Environment, 20(7): 16-37.

Fischhoff, Baruch, Slovic, Paul, and Lichtenstein, Sara, 1979, Which risks are acceptable? Environment, 21(4):17-38.

Fischhoff, B., and Whipple, C., Aug. 1980, Four conceptual approaches to health risk assessment for alternative national ambient air quality standards, Draft report prepared for U.S. EPA, Office of Air Quality Planning and Standards.

Fischhoff, B., Lichtenstein, S., Slovic, P., Keeney, R., and Derby, S., Dec. 1980, Approaches to acceptable risk: A critical guide, Oak Ridge National Laboratory, Oak Ridge, Tennessee, NUREG/CR-1614 and ORNL/Sub-7656/1.

General Accounting Office, Aug. 23, 1976, Ways to improve management of Federally funded computational systems, LCD-7511.

Green, Harold P., 1968, George Washington Law Rev., 35:1033.
Green, Harold P., 1975, The risk-benefit calculating in safety deter-
    mination, George Washington Law Rev., 43(3):791-804.
Haimes, Y. Y., Lasdon, L. S., and Wismer, D. A., 1971. On the bicri-
    terion formulation of the integrated system identification and
    system optimization, IEEE Trans. Systems, Man, Cybernetics,
    SMC-1, 296-297.
Haimes, Y. Y., and Hall, W. A., Aug. 1974, Multiobjectives in water
    resources systems analysis: The Surrogate Worth Trade-off
    method, Water Resources Res., 10(2):614-624.
Haimes, Y. Y., and Nainis, W. S., Dec. 1974, Coordination of
    regional water resource supply and demand planning models,
    Water Resources Res., 10(6):1051-1059.
Haimes, Y. Y., Hall, W. A., and Freedman, H. T., 1975, "Multiobjec-
    tive Optimization in Water Resource Systems: The Surrogate
    Worth Trade-off Method," Elsevier, The Netherlands.
Haimes, Yacov Y., 1976, Water resources systems analysis: an
    assessment of direction, in: "Economic Modeling for Water
    Policy Evaluation," R. M. Thrall et al., eds., North Holland/
    American, New York.
Haimes, Y. Y., and Hall, W. A., 1977, Sensitivity, responsivity,
    stability and irreversibility as multiple objectives in civil
    systems, Advances in Water Resources, 1(2):71-81.
Haimes, Y. Y., 1977, "Hierarchical Analyses of Water Resources
    Systems: Modeling and Optimization of Large-Scale Systems,
    McGraw-Hill, New York.
Haimes, Y. Y., and Chankong, V., 1979, Kuhn-Tucker multipliers as
    trade-offs in multiobjective decision-making analysis,
    Automatica, 15(1):59-72.
Haimes, Yacov Y., Das, Prasanta, and Sung, Kai, Sept. 1979, Level-B
    multiobjective planning for water and land, J. Water Resources
    Planning and Management Div., 105(WR2):385-401.
Haimes, Y. Y., Loparo, K. A., Olenik, S. C., and Nanda, S. K.,
    June 1980, Multiobjective Statistical Method (MSM) for interior
    drainage systems, Water Resources Res., 16(3):465-475.
Haimes, Y. Y., 1980, Hierarchical holographic modeling and overlap-
    ping coordination, Technical Report #SED-WRP-3-80.
Haimes, Y. Y., and Tarvainen, K., 1980, Hierarchical-multiobjective
    framework for large scale systems, in: "Multicriteria Analysis
    in Practice, P. Nijkamp and J. Spronk, eds., Gower Press,
    London.
Haimes, Y. Y., Craig, J. A., and Subrahmanian, J., 1980, The worth
    of streamflow data in water resources planning: computational
    results, Water Resources Res., 15(6):1335-1342.
Hall, W. A., and Haimes, Y. Y., 1976, The Surrogate Worth Trade-off
    method with multiple decision-makers, pp. 207-233, in:
    "Multiple Criteria Decision-Making: Kyoto, 1975," M. Zeleny,
    ed., Springer-Verlag, New York.

Howard, R. A., 1968, The foundations of decisions analysis, IEEE
    Trans. Systems, Man, Cybernetics, SMC-4:211-219.
Howard, R. A., 1975, Social decision analysis, Proc. IEEE, 63:
    359-371.
Kates, Robert W., 1976, Risk assessment of environmental hazard,
    SCOPE Report #8, Int. Council of Scientific Unions, Scientific
    Committee on Problems of the Environment, Paris.
Kates, Robert W., 1977, Assessing the assessors: The art and ideol-
    ogy of risk assessment, Ombio, 6(5):247-252.
Keeney, R. L., and Raiffa, H., 1976, "Decisions with Multiple Objec-
    tives: Preferences and Value Trade-offs," Wiley, New York.
Lowrance, William W., 1976, "Of Acceptable Risk," William Kaufman,
    Inc., Los Altos.
Nainis, W. S., and Haimes, Y. Y., Jan. 1975, A multilevel approach
    to planning for capacity expansion in water resource systems,
    IEEE Trans. Systems, Man, Cybernetics, SMC-5(1):53-63.
National Bureau of Standards, U.S. Department of Commerce, May 1979,
    Utility and use of large-scale mathematical models, S. I. Goss,
    ed., U.S. Printing Office, Washington, D.C.
Okrent, David, July 25, 1979, testimony presented to the forum on
    Risk/Benefit Analysis in the Legislative Process, Subcommittee
    on Science, Research and Technology, Committee on Science and
    Technology, U.S. House of Representatives.
Perspectives on Benefit/Risk Decision Making: Report of a Collo-
    quium Conducted by the Committee on Public Engineering Policy,
    1972, National Academy of Engineering, April 26-27, 1971,
    Washington, D.C.
Primack, J., 1975, Nuclear reactor safety: An introduction to
    the issues, Bull. Atomic Scientists, 31(9):15-17.
Raiffa, H., 1968, "Decision Analysis," Addison-Wesley, Reading, Ma.
Rarig, Harry, Oct. 1976, Two New Measures of Performance and Param-
    eter Sensitivity in Multiobjective Optimization Problems,
    M.S. Thesis, Department of Systems Engineering, Case Western
    Reserve University, Cleveland, Ohio.
Rarig, H. M., and Haimes, Y. Y., Jan. 1981, Model sensitivity in
    a multiobjective framework: the Dispersion Index, Technical
    Report, #SED-WRP-1-81.
Rausa, Gerald J., July 2, 1973, Some estimates of the worth of
    human life, EPA Memorandum.
Ritter, Don, July 25, 1979, testimony given at the Benefit Risk
    Forum before the House Committee on Science and Technology,
    U.S. House of Representatives.
Rose, Don, Risk of catastrophic failure of major dams, Sept. 1978,
    ASCE Journal of the Hydraulic Division, 104(HY9).
Rowe, William D., 1977a, "Anatomy of Risk," John Wiley, New York.
Rowe, William D., 1977b, George Washington Law Rev., 45(5):949.
Sage, A. P., and White, E., 1980, Methodologies for risk and hazard
    assessment: A survey and status report, IEEE Trans. Systems,
    Man, Cybernetics, SMC-10(8):425-446.

Schneier, Edward V., Jr., 1975, Legislative intelligence and the committee system, testimony before the Committee on Science and Technology, U.S. House of Representatives.

Slovic, Paul, 1978a, Judgment, choice and societal risk taking, in: K. R. Hammond, ed., "Judgment and Decision in Public Policy Formations," AAAS Selected Symposium Series, Westview Press, Boulder, Colo.

Slovic, Paul, 1978b, The psychology of protective behavior, J. Safety Res., 10(2).

Slovic, Paul, and Fischhoff, Baruch, 1979, How safe is safe enough? Determinants of perceived and acceptable risk, in: L. Gould and C. A. Walker, eds., "Too Hot to Handle: Social and Policy Issues in the Management of Radioactive Wastes," Yale University Press, New Haven.

Starr, C., 1969, Social benefit versus technological risk, Science, 165:1232-1238.

Tarvainen, K., and Haimes, Y. Y., 1980, Basic hierarchical-multiobjective optimization techniques, Case Western Reserve University, Cleveland, Ohio, Rep. No. SED-WRP-1-80.

U.S. Nuclear Regulatory Commission, Oct. 1975, Reactor Safety Study: An Assessment of Accident Risks in U.S. Commercial Nuclear Power Plants, WASH 1400 (NUREG-74/014), Washington, D.C.

U.S. Water Resources Council, Principles and Standards for Water and Related Land Resources Planning, Federal Register, September 28, 1980.

Vaupel, James W., May 1978, The prospectives for saving lives: A policy analysis, Working paper, Center for the Study of Policy Analysis, Institute of Policy Sciences and Public Affairs, Duke University, Durham, N.C.

Zeckhauser, Richard, 1975, Procedures for valuing lives, Public Policy, 23(4):419-463.

Zeckhauser, Richard, and Shepard, Donald, 1976, Where not for saving lives? Law and Contemp. Prob., 40(4):4-45.

MULTIOBJECTIVE GENERATING TECHNIQUES FOR RISK/BENEFIT ANALYSIS

Jared L. Cohon,* Charles S. ReVelle*
and Richard N. Palmer**

*Department of Geography   **Department of Civil
   and Environmental Engr.   Engineering
The Johns Hopkins University   University of Washington
Baltimore, Maryland   Seattle, Washington

The point of departure for this paper is the common methodological and policy-making characteristics shared by problems in "risk/benefit analysis" and multiobjective analysis. Indeed, we are not sure of the differences between the two problem types: risk/benefit analysis seems to be simply a new phrase for a specific setting of multiobjective analysis. Multiobjective problems have many conflicting objectives, presenting a challenge to decision makers to select an alternative that will somehow balance the trade-offs among the objectives. Similarly, risk/benefit analysis is directed at problems in which a similar compromise must be struck between the benefits of an action and the risks posed by or inherent in that action. Our claim is that risk/benefit analysis is just multiobjective analysis over two specific objectives: benefits (itself a multidimensional measure of system performance in the general case) and risk.

Given the similarity in decision-making characteristics of risk/benefit and multiobjective analysis, the well-developed methodologies for the latter should be applicable to the former. Multiobjective analysis is briefly reviewed below, after which we propose a certain class of multiobjective techniques--generating methods--for risk/benefit analysis. The generating methods exhibit features that are particularly compatible with the complex, politically sensitive problems that are typical of situations that arise in risk/benefit analysis.

An example is presented to demonstrate the basic notions. Our conclusions focus on some challenges to risk/benefit analysts.

MULTIOBJECTIVE ANALYSIS

In this section and the one that follows multiobjective solution methods are briefly reviewed. Much of this material is based on Cohon and Marks [1975] and Cohon [1978].

A decision-making problem is said to be multiobjective if there are many objectives which conflict in the sense that the decision which is best for one objective is different from the plans which optimize the other objectives. Thus, almost all public decision making problems, with the many interest and interested groups affected by them, are multiobjective. Even our private decisions tend to be multiobjective in nature. Which car should you buy? The cheapest car may be too small for your purposes while the largest car may be too expensive. To make a decision, you must evaluate the trade-off between cost and size and strike a compromise, no matter how unsatisfying that compromise may be.

A useful conceptual representation of a multiobjective problem is shown in Figure 1. The axes of the figure are two objectives, Z_1 and Z_2, both of which are to be maximized. The space represented is called the objective space. Constraints on the problem define the set of feasible decisions, the feasible region in objective space. Using the notion of dominance, we can eliminate many feasible decisions as candidates for the best compromise solution (BCS)--that solution selected for implementation by decision makers.

For example, alternative C in Figure 1 can be eliminated from further consideration because there are other feasible solutions which dominate it. Point D, which gives more of Z_2 and the same amount of Z_1, is clearly a better choice. Any solution in the area to the northeast of C dominates C. Point C is said to be inferior.

Solution D in Figure 1, on the other hand, is not dominated. It is said to be nondominated or noninferior because there are no feasible solutions that dominate it. The cross-hatched portion of the boundary of the feasible region, from A to B, is the noninferior set, the collection of decisions that cannot be clearly eliminated from further consideration.

Notice that as we move along the noninferior set from A to B we are trading off Z_2 to gain more of Z_1. This presence of trade-offs among solutions is an essential feature of the noninferior set.

We've come quite far in narrowing the decision to the noninferior set. But, which of the infinity of decisions in the noninferior set should be selected; which is the BCS? It is on the answer to

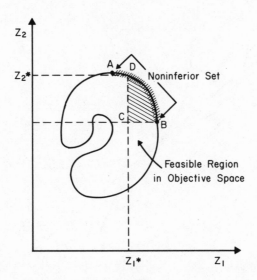

Fig. 1.   The noninferior set for an arbitrary feasible region in
          objective space.

this question that the distinction among multiobjective techniques
is drawn. Generating methods, the techniques that we promote in
this paper, are based on unstructured decision-making from this
point on. The rationale is that decision makers need to be informed
and the set of noninferior decisions does this quite nicely. The
very personal process of striking a compromise among the competing
objectives is left to decision makers.

Preference-oriented methods, by contrast, construct a formal
process which leads the decision maker to a BCS. The key is the
quantification of preferences, and the manner in which preferences
are represented distinguishes one preference-oriented technique
from another.

In its most general form, a preference representation can be
thought of as a utility function. A noninferior set is presented
in Figure 2 along with indifference curves which are loci of equal
utility. Given these indifference curves, theory shows that the
point at which an indifference curve is tangent to the noninferior
set is the BCS--the alternative which maximizes utility.

A familiar representation of preferences is a set of weights
on objectives. Weighting objectives is equivalent to constructing

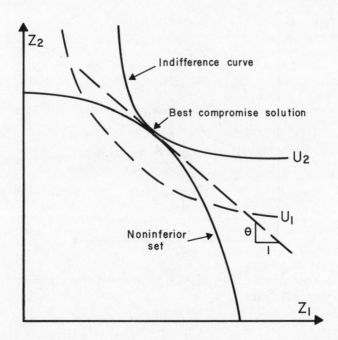

Fig. 2. Geometry of noninferior set, indifference curves, and best
        compromise solution

a very special utility function which yields linear indifference
curves. The dashed line in Figure 2 with slope θ (which is negative) corresponds to a weight on Z_2 of 1 and a weight on Z_1 of
$(-\theta)$.

Whatever form preferences take, the underlying philosophy of
preference-oriented techniques is the same. In the next section,
we will argue against this philosophy and, in the process, for the
use of generating methods in public sector problems in general and
water resources planning problems in particular.

GENERATING METHODS VS. PREFERENCE-ORIENTED TECHNIQUES

Our basis for objecting to preference-oriented methods is twofold: the nature of public decision making processes makes the
techniques inapplicable or, at best, cumbersome; and, the public
decision maker may not be ready for such methods.

Public decision-making processes are not well understood. We
do know, however, that they are dynamic and usually characterized
by many decision makers. Furthermore, all of the decision makers
may be unknown to the analyst, and, if they are known, they may be
relatively inaccessible. This raises the following questions:
Whose preferences should be used? How should differences among many
sets of preferences be recognized? How should unknown (or unknow-
able) preferences be accounted for? How should changing preferences
be represented? These are difficult questions that have not been
answered. And, it seems to us, they must be answered before preference-oriented techniques can be used to support real public decisions.

Notice that generating techniques tend to escape this tangle
of imponderables. No matter what form the decision-making process
may take, the noninferior set as a representation of the range of
choice is valuable information. We recommend, in effect, a retreat
to the position that political decision making is inherently too
poorly defined to yield to mathematical formalism; as analysts, our
most productive role is as information providers. However, there
is an important qualification: any successful analysis must be
sensitive to the nature of the decision-making process. Thus, objectives, decision variables and constraints, if they are to represent
the real problem at hand, must be formulated from some understanding
of what is important to decision makers.

Our second argument is based on decision makers' reluctance to
provide explicit statements of preference. This is especially true
in cases typical of risk/benefit analysis: expected lives saved
vs. cost for flood control; expected new cancers vs. the benefits
of a chemical; and expected cancer deaths vs. barrels of oil saved
by a new nuclear power plant. Should we really expect decision
makers to tell us explicitly (before the decision implications are
understood) that the saving of a life is worth $279,000, but no
more? We think not.

Generating methods do not avoid the making of value judgments;
to do this would be to avoid decisions. Indeed, the trade-offs
involved are clearly presented to decision makers, so the value
implications are apparent. It is then up to decision makers to do
as they see fit (with the help of the analysts), a procedure we
find to be more compatible with the way public decisions are made.

AN APPLICATION TO WATER SUPPLY PLANNING

The following application demonstrates a "risk/benefit" analysis
using a multiobjective generating technique. The material is derived
from Palmer et al. [1980a] and Palmer et al. [1980b].

The Problem

The proper management and operation of reservoirs used for
water supply has been historically one of the major research topics
considered by water resources engineers and analysts. Interest in
this topic has been accentuated in recent years by several factors.
Many urban areas have begun to find their water supply systems inadequate in the face of continued population growth and increased per
capita water consumption. This problem is often compounded by a
paucity of new sources of water. Many conventional means of expanding water supply (such as the construction of new reservoirs, heavy
reliance on groundwater and interbasin transfers) have come under
heated attack. These attacks in many cases have decreased the possibility of conventional approaches serving as politically feasible
solutions. Thus, agencies charged with developing adequate water
supplies have begun to explore the potential of more efficient management techniques in extending the life of their water supply.
This situation has been quite graphically illustrated in the case
of the water supply system of the Washington, D.C. metropolitan
area (WMA).

The WMA is composed of the District of Columbia (D.C.) and the
surrounding areas in the states of Maryland and Virginia. Drinking
water for this region is supplied by three major water agencies,
the Washington Aqueduct Division (WAD) of the U.S. Army Corps of
Engineers, the Washington Suburban Santiary Commission (WSSC), and
the Fairfax County Water Authority (FCWA) (Figure 3). WAD obtains
all of its water from the Potomac River near Little Falls, Maryland.
Currently, all of the water used by the Fairfax County Water Authority is drafted from a 9.8 billion gallon reservoir located on
Occoquan Creek. The Washington Suburban Sanitary Commission obtains
water from the Potomac River and from two reservoirs in series on
the Patuxent River. Because these reservoirs are in series and are
located close to one another, in this analysis they were considered
to be one reservoir, referred to as the Patuxent Reservoir.

All of the water which supplies the WMA originates in the
Potomac and Patuxent River basins. The Potomac and Patuxent basins
drain approximately 15,000 square miles including portions of Maryland, Virginia, Pennsylvania, West Virginia and the whole of the
District of Columbia (Figure 3). The Potomac basin, the larger of
the two, extends 380 miles from the southwesternmost tip of the
state of Maryland down to the Chesapeake Bay. The area is generally
humid with average annual rainfall ranging from 32-50 inches per
year.

There are two points of large water demand in the basins. The
largest demand for water occurs in the WMA where water must be supplied to nearly three million residents. Water is also needed to
dilute the regional discharge of domestic sewage which flows into

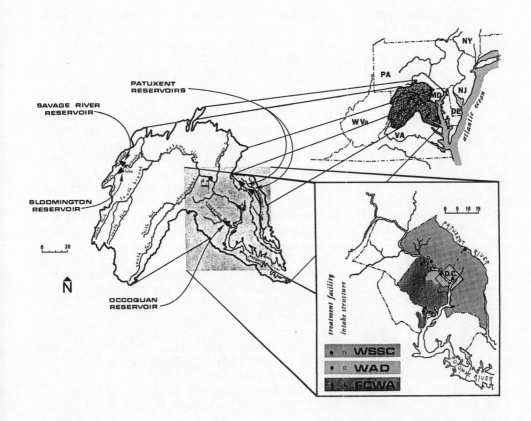

Fig. 3. Washington, D.C. metropolitan area water supply system

the Potomac Estuary. The quantity of water which is reserved for
this purpose is referred to as environmental flow-by. A second area
of large water demands exists at Luke, Maryland, which is located on
the North Branch of the Potomac River. A large pulp paper mill industry is located at Luke and water is used for processing, cooling and
waste dilution. It is beneficial to this area that a large, dependable supply of water be made available from the Savage River Reservoir and from the soon to be completed Bloomington Reservoir.

Up to the present time, no drought has resulted in large water
supply deficits in the WMA. Two especially severe droughts have,
however, occurred in the past fifty years, one during the early
1930s and a second during the mid-1960s. One feature of both of
these droughts was the recurrence of low flows over a period of three
to four consecutive years.

During the 1930s drought, no major reservoirs existed to augment
low flows. In response to that drought and to increasing water use
in the Potomac basin a number of reservoirs were constructed. In
1942 Triadelphia was built to aid in the supply of water from the
Patuxent River. Twelve years later, in 1954, a second reservoir
was built on the Patuxent. In 1950, Savage River Reservoir was built
for flood control purposes and to help supply water to upstream users
at Luke, Maryland. In 1957 Occoquan Reservoir was built to aid the
water supply of the Virginia suburbs. These three downstream reservoirs left only those areas served by the Corps of Engineers without
a firm source of water. In 1956, 25 years after the drought of the
thirties, the Corps of Engineers began a detailed study of the WMA
water supply. Made public in 1963 and entitled the Potomac River
Basin Report, the North Atlantic Division of the Corps of Engineers
proposed a guide for water resources development in the basin. The
plan included 16 major reservoirs with over 500 billion gallons of
storage for water supply and low flow augmentation and another 200
billion gallons of storage for flood control. This "structural"
solution received an extremely negative public response. Currently
only one of the original sixteen reservoirs (Bloomington Reservoir)
has made it through the intricate political maze to the point of
construction.

If the attitude of the citizenry of the Potomac basin has
changed perceptively since the release of Potomac River Basin Report, it has become even more firmly opposed to the construction
of new reservoirs. There is little chance that any of the remaining fifteen projects originally suggested by the Corps will ever be
constructed. Consequently, questions concerning the location, size
or sequencing of new reservoirs are irrelevant. Instead, the goal
of current planning is to determine how long the life of this urban
water supply could be extended with proper management of the already
existing reservoirs under a variety of management constraints and
objectives.

Analysis

An optimization model, called the Linear System Yield Determination (L-SYD) model, was formulated to represent the major physical components of the WMA water supply system. One objective of the model was to meet water demands as far into the future as possible, referred to as the "life" of the system. In effect, this is a risk measure: "Life" was defined as the number of years in the future in which the most severe drought on record (that of the 1930s) could still be met, even with growing demands.

The WMA system life objective was traded off against two other objectives: upstream demands for water and environmental flow-by. The trade-off curve relating the life of the WMA to upstream release requirements is given in Figure 4 (for an environmental flow-by of 100 mgd). This figure indicates that the upstream release requirement can be maintained as high as 174 mgd without adversely affect-

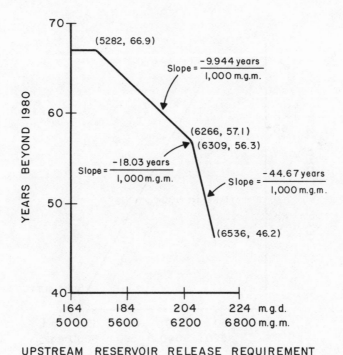

UPSTREAM RESERVOIR RELEASE REQUIREMENT

Fig. 4. Trade-off curve for upstream reservoir release requirement and system yield

ing the life of the WMA system. As the upstream release requirement
increases beyond this value, the yield begins to decrease.

This trade-off curve offers valuable insight into the effects
of upstream release requirements on water availability downstream
during prolonged droughts such as that which occurred during the
1930s. Water must be available for both upstream releases and downstream flow augmentation. As the upstream release requirement is
increased, water that could otherwise be used to augment downstream
flow must be reserved for the upstream release requirement. The
trade-off curve presented in Figure 4 gives the water resources managers (decision makers) in the region some indication of the impact
that various upstream release requirements might have on the life
of the WMA water supply.

The second trade-off of interest is that between the life of
the WMA water supply and the environmental flow-by. Figure 5 presents the results of the trade-off between environmental flow-by
and the life of the WMA water supply. From the figure it can be

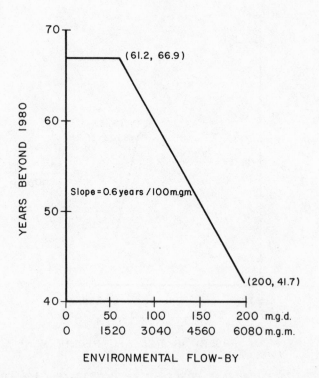

Fig. 5. Trade-off curve for environmental flow-by and system yield

seen that an environmental flow-by of 61.2 mgd can be sustained with no effect on the life of the system. If the environmental flow-by is increased beyond this quantity, however, the life of the system begins to decrease at a constant rate.

This analysis supplies valuable information to the decision-makers in their attempt to establish reasonable values for environmental flow-by. The study indicates that there does exist some maximum environmental flow-by beyond which the life of the system decreases significantly. Likewise, it indicates that the environmental flow-by need not necessarily be set to zero for fear of extreme negative impacts on the downstream water supply. Rather the results generated from a severe, prolonged drought indicate that the value can be set at 61.2 mgd without suffering any adverse consequences.

THE CHALLENGE: IDENTIFICATION AND QUANTIFICATION OF OBJECTIVES

We have shown that multiobjective analysis, particularly generating methods, provide a useful methodological framework for risk/benefit analysis. We see no serious barriers to this, save one: the identification and quantification of risk measures and other objectives.

Our system "life" objective in the previous section is a relatively crude, yet creative measure of the system performance that has an implicit risk component, specifically, the risk associated with droughts larger than the worst drought on record. We would not claim that this is an adequate, explicit risk objective of the sort that we need to support decisions.

Exploratory ideas on risk objectives were advanced by Haimes, et al. [1975]. Current, but yet to be published work, has been performed at Harvard and Cornell, which we think will lend new insights and define new challenges in this most important area. We feel that the area of risk measurement in a manner that is meaningful to decision makers is a crucial area that deserves further research.

REFERENCES

Cohon, J. L, 1978, "Multiobjective Programming and Planning," Academic Press, New York.
Cohon, J., and Marks, D., 1975, "Review and evaluation of multiobjective solution techniques," Water Resources Res., 11.
Haimes, Y., Hall, W., and Freedman, H., 1975, "Multiobjective Optimization in Water Resources Systems: The Surrogate Worth Trade-off Method," Elsevier, Amsterdam.

Palmer, R. N., Smith, J. A., Cohon, J. L., and ReVelle, C. S.,
     1980a, Reservoir management in the Potomac River Basin, sub-
     mitted for publication.
Palmer, R. N., Wright, J. R., Smith, J. A., Cohon, J. L., and
     ReVelle, C. S., 1980b, "Policy Analysis of Reservoir Operation
     in the Potomac River Basin," 3 vols. (see especially Vol. 2),
     Water Resources Research Center, University of Maryland,
     College Park, Md.

THE RISKS OF BENEFIT-COST-RISK ANALYSIS

Ronald A. Howard

Department of Engineering-Economic Systems
Stanford University
Stanford, California

INTRODUCTION

First of all, it is a pleasure to be here to talk to you and
to have an opportunity to deal with an audience of people that normally I would not see. Five years ago, we had a conference in this
very room on questions of risk assessment. At that time I talked
about some early ideas I had on how to approach the question of the
value of life.

Since then, there is some good news and some bad news. The
good news is that I think I understand the problem well enough to
advise individuals in making risky decisions that affect their
lives. The bad news is that the very concepts that make it possible to make such decisions for an individual virtually preclude
"public" decisions about risk. For me, this is a happy result,
for rather than, as someone said earlier, bending logic and ethical considerations to fit public decision making, I would prefer
to bend public decision making to fit logical and ethical considerations.

Since many of you may not know the kind of approach that I
have been working on, I will divide my time between presenting an
outline of that approach and discussing an aspect of it that I have
not previously presented. Finally, I shall present some cautionary
observations on the use of benefit-cost-risk analysis for public
decisions.

A CONCEPTUAL MODEL FOR LIFE AND DEATH DECISIONS

The key question is whether one would take on risk for material benefits, let us say, for money, although we realize there are
other things besides money that might encourage us to take on risks.
In particular, would I take on an additional risk with probability
p of killing me for some amount of money x? Figure 1 shows what
this means in terms of the choices I have to make. I am sitting
with some wealth level and looking at this decision prospect. If I
reject this opportunity, then I have (and this is very important)
a future life lottery; that is, I am not entering upon a certainty
in that case but rather a very uncertain prospect--the way the rest
of my life will be lived starting with my present wealth. If I
accept this deal, on the other hand, then I might die with probability p having received the additional payment x. If I live, I
start my future life lottery in a better materialistic situation,
since I shall have received the payment x. In real life, we encounter deals like this when, for example, we travel on business.
How large must x be to induce me to undertake an incremental
risk of death p?

Well, when I think about how x is related to p in qualitative terms, I find, as shown at the top of Figure 2, that the bigger p is the bigger x is going to be--and this is no surprise.
But I also realize that for me there is some maximum probability
of dying that you can induce me into taking for money, and I call
that p_{max}. Another interesting aspect is that I can calculate
what value of life v would give me a curve like this in an
expected value sense. Since the payment x would be the value of
life v times the probability p, then the value of life should
be x divided by p. If we show on the same scale x divided by

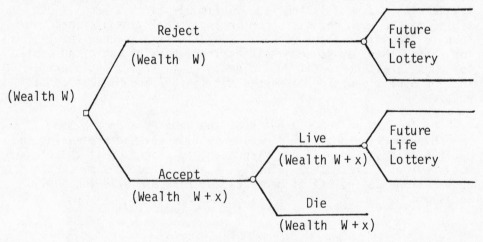

Fig. 1. A risky choice

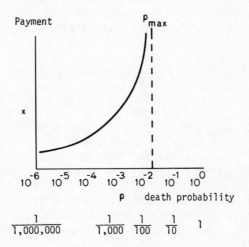

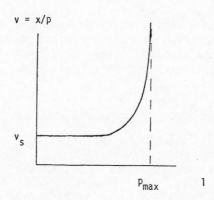

Fig. 2. Qualitative forms

p we get the curve at the bottom of Figure 2. This value of life becomes infinite as p approaches p_{max}, but for a very large range of p and certainly for the safety range, which I define as one in one thousand chance of dying per year or less, the value of life is relatively constant. On the basis of these qualitative considerations, I have developed quantitative models, and I shall present some of their results in brief, but I want to emphasize another aspect of the work; namely, what you have to believe to use such models for making risky decisions.

Beliefs Supporting the Quantitative Analysis

When is a person being logically consistent when he says, "I will not trade my life for money"? Is that a reasonable position? Terminology and notation I shall use are shown in Table 1. We need a surrogate for the money to be received in each year of life. I use what I call constant annual consumption, which is the annual amount of money in real dollars (without inflation) that you feel is equivalent to what you actually will get for the rest of your life. It is as if you sold out all your present earning ability, but were still going to do the work in exchange for some constant real consumption. Incidentally, consumption is not meant literally --you may choose to give the money away.

As for the quantity of life, we let ℓ be the number of years of life remaining. So now we are measuring in an approximate sense both the quality and quantity of life. I might add parenthetically that we are now interested in extending this model to include what we call health states or life states. For example, some people may think there are fates worse than death, like being a quadriplegic.

Table 1. Notation

c = constant annual consumption

ℓ = number of years of life remaining

$u(c,\ell)$ = utility of c,ℓ

p = incremental probability of death

x = payment to undertake risk

ζ = amount of annual consumption for rest of life that \$1 will buy

Let $u(c,\ell)$ represent the utility to the individual of that consumption-lifetime pair. I am using utility here in the von Neumann-Morgenstern sense which means that it represents not only the relative value of c and ℓ pairs but also my risk preference for uncertain combinations of c and ℓ .

Continuing in Table 1, p is the incremental probability of death and x is the payment to be received if you take the risk. We must specify what you will do with that payment. We assume that you will use it to buy an annuity to increase your constant annual consumption. The amount of annual consumption for the rest of your life that $1 will buy we define as ζ . Therefore, the payment x will increase constant annual consumption by ζx .

Turning to Figure 3, we can translate the original tree struc- ture into our new terms, focusing on the remaining life probability distribution $\{\ell\}$ and ignoring uncertainty in consumption for con- venience. If you reject the proposition, you receive $u(c,\ell)$ with probability $\{\ell\}$. If you accept the proposition and you die, you end up with utility $u(0,0)$. If you live, you are paid x which you use to convert to an amount ζx of annual consumption dollars. You thus receive the utility $u(c + \zeta \hat{x},\ell)$ with the same lifetime probability distribution $\{\ell\}$ you would have had before.

Now in order to be indifferent between these two alternatives and hence establish what is the breakeven x for a given p, the expected utility of both alternatives must be equal. Thus,

$$\langle u(c,\ell)\rangle = p\, u(0,0) + (1-p) \ \langle u(c + \zeta x,\ell)\rangle \tag{1}$$

where $\langle \ \rangle$ denotes expectation.

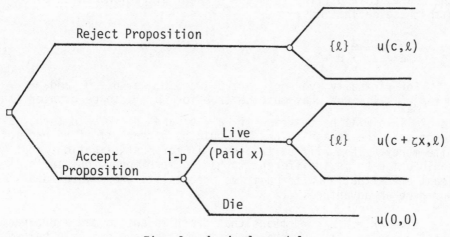

Fig. 3. A simple model

Solving for p, we have

$$p = \frac{<u(c + \zeta x, \ell)> - <u(c,\ell)>}{<u(c + \zeta x,\ell)> - u(0,0)} , \tag{2}$$

which allows one to solve for p once x is specified and hence develop the x versus p curve of Figure 2. Note that as x approaches infinity, p approaches a p_{max} given by

$$p_{max} = \frac{<u(\infty,\ell)> - <u(c,\ell)>}{<u(\infty,\ell)> - u(0,0)} . \tag{3}$$

Let us now determine when a person has a p_{max} strictly greater than zero and strictly less than 1: $0 < p_{max} < 1$. Such a person must be willing to undertake a risk for money, provided that the risk is sufficiently small, and yet unwilling to accept any amount of money to accept a large enough risk of death.

We begin by defining the numerator of Equation (3) to be a quantity we shall call the "advantage," a,

$$a = <u(\infty,\ell)> - <u(c,\ell)> . \tag{4}$$

This is the increase in utility over the present situation as a result of having limitless money. Similarly, we define the disadvantage, d, as

$$d = <u(c,\ell)> - u(0,0) , \tag{5}$$

the decrease in utility from the present situation caused by dying. With these definitions,

$$p_{max} = \frac{a}{a + d} . \tag{6}$$

For virtually everyone $a \geq 0$, $d \geq 0$, $a + d > 0$ and $0 \leq p_{max} \leq 1$. But what must be true for p_{max} to be greater than zero? This will be the case if $a > 0$ and $d < \infty$. In turn, $a > 0$ means that the person thinks he would be better off if he had infinite money. This will be true even if for most situations that could arise he is indifferent to having more money, but there is at least one situation with positive probability where more money would be an advantage.

Similarly, $d < \infty$ means that death is not so bad a prospect that no action with any chance of it must be avoided. To believe the contrary would leave one paralyzed in the face of life. Consequently, for almost everyone, $p_{max} > 0$.

When will p_{max} be less than 1? When $d > 0$ and $a < \infty$. A person will have $d > 0$ when he believes he has something to lose by dying; this will be the usual case except for people on the brink of suicide. A person will have $a < \infty$ if he believes that having infinite money will not bring him infinite joy. Otherwise, he would take any action that had even a tiny chance of boundless wealth. Therefore, for almost everyone, $p_{max} < 1$.

We can conclude from $p_{max} > 0$ that there are some life risks so small that money would induce a person to assume them, and from $p_{max} < 1$ that there are some life risks so large that no amount of money could induce the person to assume them. Consequently, virtually no one can say

a) that he does not or would not risk his life for money

and

b) that he will accept any risk for enough money.

A Simple Example

These conclusions are correct for a wide range of utility functions. However, to show how a simple model can be used to produce practical results, let us examine a typical example.[1] To specify the model, we must first assign a probability distribution to remaining life; we shall use a standard mortality table and assume that the person is a 25-year-old male. We must also specify his opportunity to turn present funds into future consumption; we shall assume that he is able to obtain a 5% interest rate on real dollars. He has a constant annual consumption of $20,000 per year. We establish his risk preference by determining that he has an exponential utility function on money and that he is indifferent between his present situation and a proposition that would give him a 82% chance of $40,000 annual consumption for the rest of his life or an 18% chance of $10,000 annual consumption for the rest of his life. Finally, we learn from him that he would give up 10% of his income for a 5% increase in his life.

---

[1] Details are presented in "Life and Death Decision Analysis," Research Report EES DA-79-2, Department of Engineering-Economic Systems, Stanford University, December 1979.

These specifics are sufficient to compute the x versus p
curve for the individual as shown in Figure 4. We observe that
p_{max} for this person is about 0.103: he would not accept a higher
probability of death for any amount of money. The value of life
curve at the bottom of the figure shows that in the safety region
the curve is virtually constant and equal to a small-risk life
value of about $2.43 million. Compare this with the economic value
of life that would be obtained by determining the cash amount that
would allow this person's annual consumption; this would be only
$363,000, less than one-sixth the small-risk life value.

One difficulty I have encountered is that people often forget
the details of the story I have told and remember only a $2.43 mil-
lion value of life. Notice that it is obviously wrong in terms of
the model to conclude that you could kill this person for $2.43
million. To minimize the possibility of such erroneous conclusions,
I have begun to use the term "micromort" for a one-in-one million
chance of death, with symbol µmt. Then I can say that the person
values his life at $2.43 per µmt and simply add that this price is
only good for the first 1000 or so micromorts presented to him; if
you try to buy a million, he will not sell.

In my opinion, the idea of a value per micromort is more fundamental than any model that might be used to help establish it. A
person could simply choose a value per micromort and then use it to
see if the resulting decisions felt comfortable; if not, he could
change it until he was happy with the result.

We illustrate the consequences of having a value of $2.43 per
micromort by using U.S. accident statistics as presented in Table 2.
This table shows the various types of accidents, the number of
annual micromorts they contribute, and the value of these micromorts at $2.43 each. For example, the individual[2] would pay $656
to be totally freed from the risk of motor vehicle accidents. (Of
course, we must require that he forget that he made this deal so
that his behavior will stay the same.) Note that he would pay about
$1000 per year to be free of just the first three types of accidents.

To summarize, virtually everyone must be willing to exchange
risks to life for material benefits at small levels of risk. The
value per micromort can be set by models or assigned directly and
used for just such decisions.

---

[2]We assume that the individual would assign these probabilities as
his own risks. In most cases, he will wish to modify them to suit
his style of life.

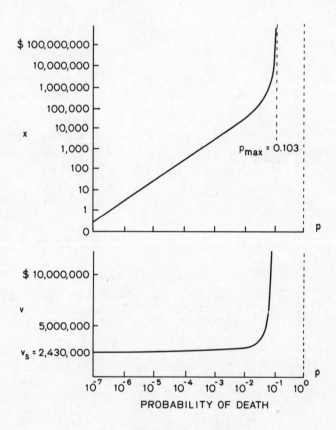

Fig. 4.

Table 2. U.S. Accident Death Statistics

Type of Accident	Total Annual Deaths	Probability of Death in Micromorts/Year (1 Micromort [µmt] = 10⁻⁶)	Payment of Base Case Individual to Avoid Hazard @ $2.43/µmt
Motor vehicle	53,041	270	$ 656.00
Falls	20,066	100	243.00
Fire and Explosion	8,084	40	97.20
Drowning	5,687	28	68.00
Firearms	2,558	13	31.60
Poisoning (solids and liquids)	2,283	11	26.70
Machinery	2,070	10	24.30
Poisoning (gases and vapors)	1,648	8.2	19.90
Water transport	1,630	8.1	19.70
Aircraft	1,510	7.5	18.30
Inhalation and ingestion of food	1,464	7.3	17.70
Blow from falling or projected object or missile	1,459	7.3	17.70
Mechanical suffocation	1,263	6.3	15.30
Foreign body entering orifice other than mouth	1,131	5.7	13.90
Accident in therapeutic procedures	1,087	5.5	13.40
Railway accident (except motor vehicles)	1,027	5.1	12.40
Electric current	1,026	5.1	12.40
Other and unspecified	6,163	31.0	76.50
TOTAL	113,563	580.	$ 1,384.00

Virtually no one will take a sufficiently large risk for any
amount of money: you can never buy 1,000,000 micromorts from anyone. In fact, we use models to compute p_{max}, the death probability that a person would never exceed no matter what he was paid.

THE FLAWS OF BENEFIT-COST-RISK ANALYSIS

Now that I have presented some of my views on how to help
people make risky decisions, let me turn to the question of why I
do not think these methods are appropriate for public decisions.

Technical Flaws--Benefits and Costs

The first difficulties I see are technical flaws. These flaws
are present not only in the aspect of risk but throughout benefit-
cost-risk analysis. We are all familiar with the problems of quantifying benefit, and they are indeed great. We must ask, "Benefits
to whom, and valued by whom?" But there are equally great problems
surrounding cost. We must ask not only who is going to pay the
costs but also how will those costs be measured. In a world of
extensive government intervention, this job is even more difficult
than is usually appreciated. For example, in determining the cost
of a dam, do you use the price of so-called "dumped steel" from
Japan or the price of steel made in Pittsburgh by U.S. Steel?
These costs are not so much economic questions as political questions, since the government, by changing the rules, has a major
effect on the price. Similarly, if federal law requires that union
labor be used on a construction job, do you use these union labor
rates in computing the costs even when there are qualified people
willing to work for less? Taxes pose a major problem. In comparing private and public projects, do we include taxes as costs in
the private calculation but not in the public calculation? Think
of the importance of this issue in deciding whether a postal service ought to be conducted by business or government. If you
assume that all taxes paid by private enterprises are costs to
them but not costs to a government enterprise, it will be difficult to escape the conclusion that virtually all activities should
be conducted by the government. And yet the postal case makes us
question that conclusion.

An important issue that is clearly related to water development is the use of federal land. If a project will utilize many
acres of land currently held by the government, what cost should
be assigned to this use? Is it the price that private entrepreneurs would pay for that land for residential, commercial, or
agricultural uses? Or do you say, "This is government land, so it
is free."

In a society with something like a 40% government economy, these issues are no longer peripheral. We must understand that the cost side of economic analysis has as many difficulties as the benefit side. It is interesting to note that, in the early years of the Soviet Union, there was a real question of whether a purely socialist economy could function since its planners would not have available, from any free markets, the pricing signals that a free market provides. Fortunately for those planners and fortunately for us, there is a sufficient set of free markets in the world to provide those signals. However, if our economy continues its present trend toward government intervention in markets, we shall increasingly encounter the problem of knowing the real cost of anything.

We should note that this problem is a problem only for a government decision-maker. The private entrepreneur must make the best decision he can, given the costs he faces. It does not matter to his calculation whether the high cost of bread is a consequence of drought or of government policy. However, since the government can change apparent costs by its own decisions, it must be concerned with this effect in assessing costs. For example, the government could probably show the great public benefit of a solar power program once it had imposed the equivalent of $100 per bbl tax on all fossil fuels. The problems of benefit-cost analysis that I have discussed only arise when dealing with problems from a government perspective.

Technical Flaws-Risk

The problems of benefit-cost analysis are compounded by the addition of risk. We must ask: risks to whom and assessed by whom? Who is going to evaluate the risk; that is, who is going to make the many value judgments surrounding risk taking that I described earlier? Someone will have to specify attitude toward risk taking, time preference, trade-offs between quantity and quality of life, etc. Even if I were ethically happy about the idea of someone making such judgments for others, the practical problems are overwhelming.

Ethical Flaws

But I am not happy about the idea of someone making a decision for another without his consent; I consider it unethical. If benefit-cost-risk analysis is used as a means to justify pushing peaceful, honest people around, then I must add ethical problems to its technical problems.

The basis for the use of benefit-cost-risk analysis is usually that if you can show (which I doubt, technically) that for some program the net benefit summed over all individuals would be positive, then the program is a good one whether or not compensation to those hurt is actually paid. I do not find this argument at all persuasive, and I doubt that many other people would, too. If I tell you that the freeway constructed right beside your house is a good idea because it had a positive net benefit, you mak ask, "That's fine, but where's mine?", if you can hear me over the noise. If the government takes your property wholly or partially without compensation that you willingly accept, that to many is unethical. But suppose that you do willingly accept the compensation? Well then you do not need government action because entrepreneurs will be happy to make the deal. So the only time you could justify government action, you do not need it. And, of course, as the ethical issues disappear, so do the technical issues, for each person can assess his own benefits, costs, and risks. Now that is the world I wish for.

REFERENCES

Howard, Ronald A., 1980, On making life and death decisions, in "Societal Risk Assessment: How Safe is Safe Enough?" Richard C. Schwing and Walter A. Albers, Jr., eds., Plenum Press, New York.
Howard, Ronald A., 1980, An assessment of decision analysis, Special Issue on Decision Analysis, Operations Res., 28, No. 1.
Howard, Ronald A., 1979, "Life and Death Decision Analysis," Research Report EES DA-79-2, Department of Engineering-Economic Systems, Stanford University.
Howard, Ronald A., 1978, Life and death decision analysis, Proc. 2nd Lawrence Symposium on Systems and Decision Sci., Berkeley, Cal.
Howard, Ronald A., Matheson, James E., Owen, Daniel L., 1978, The value of life and nuclear design, Proc. Am. Nucl. Soc. Conf. on Probabilistic Analysis of Nucl. Reactor Safety, Los Angeles.

METHODS FOR DETERMINING THE VALUE OF MODEL DEVELOPMENT IN COST/BENEFIT/RISK ANALYSIS

W. Scott Nainis

Arthur D. Little, Inc.
Cambridge, Massachusetts

INTRODUCTION

Typically when an analysis of some combination of costs, benefits, and inherent risks of a set of alternatives is to be assessed,
it is often decided that some effort in model development and associated data collection should be undertaken. Often the decision is
made to go ahead with a sizeable modeling and data collection
effort. In this paper I discuss the issues which affect that decision to use models in cost/benefit/risk analysis and discuss these
issues with respect to on-going model development activity concerned
with providing decision support for toxic waste decontamination/
containment planning.

SETTING THE OBJECTIVES AND SCOPE OF A COST/BENEFIT/RISK ANALYSIS

An important issue in considering the value of a model in cost/
benefit/risk (C/B/R) analysis is how the organizational entity performing the C/B/R analysis is postured with respect to the outcomes
and associated objectives resulting from the possible range of decision alternatives. The obvious example of the importance of this
concern is the difference between public sector and private sector
objectives in C/B/R analysis.

A good illustrative example is toxic substances handling. Private corporations have a responsibility for insuring the safety of
operations for their employees and to the public at large. Unfortunately, many of the key issues associated with toxic substances
are clouded with high degrees of uncertainty. Difficult questions
abound. Is the compound toxic to humans at various concentration

levels? Has a compound been released to the environment? Has the compound migrated away from containment areas and off the plant site? How far should an organization go in considering these questions? What level of knowledge and certainty should be developed? Government agencies concerned with regulation of toxic substances may have a substantially different opinion than do the firm's stockholders. One intent of environmental regulatory programs (e.g., Toxic Substances Control Act) is to reduce the nature of the gap between private sector and public sector goals and priorities.

The difference between priorities and objectives among various groups viewing a decision is not limited to those found between the public and private sector. Within public agencies, varying points of view will occur. A useful relevant example is the breakdown of responsibility with the U.S. Army. The Toxic and Hazardous Materials Agency is concerned with the fate and control of toxic and hazardous materials; however, the U.S. Army Surgeon General is responsible for setting the allowable standards for toxic and hazardous material contamination. The result of this delegation of responsibility is that, in the overall picture, it may be difficult to assess and to act in order to balance overall costs, benefits, and their intrinsic risk. If, for example, two contaminants were found to be 100% over standard, it may not be of equal benefit in terms of risk reduction to reduce each level of contamination back to its stated standard.

Analysis of toxic substances risk levels must be taken into account in setting standards. The limitation with standards, of course, is that by their nature they incorporate an average assumed exposure and associated risk. In any specific situation (i.e., plant site) the actual human life and environmental impact risks can vary widely depending upon the likelihood of toxic material exposure involved in that site.

The development of objectives and the assembling of information in order to determine methods for meeting those objectives is really the task of a data collection and modeling effort in a cost/benefit/risk context. One major point is clear: a data collection and modeling effort must address as explicitly as possible those indicators which represent cost and benefits along with their associated risks.

DETERMINING THE VALUE OF A MODELING EFFORT

The value of data collection and associated modeling effort depend upon how well decisions are improved overall through use of the model. Formal approaches [Nickerson and Boyd, 1980] utilize a Bayesian statistical approach where the quality of decision making is probabilistically estimated assuming an <u>a priori</u> and <u>a posteriori</u> distribution on the results of a set of <u>defined</u> alternatives. In general terms, the approach to evaluating the results of a modeling effort can be expressed mathematically.

For example, in an environmentally oriented program such as toxic materials containment and decontamination, the intended objective would be to "reduce the minimized 'expected regret' with the use of the modeling activity." We will refer to the general class of data collection, handling, and modeling as constituting a decision support system (DSS) [Keen and Scott-Morton, 1978].

Consider a decision space definable without a DSS as A_M (Manual). The alternative space definable with a DSS is likely to be expanded to A_{DSS}. Thus,

$$R(A_{DSS}) \geq R(A_M)$$

where $R(\ )$ denotes the "power" or size of the decision space.

The minimum "expected regret" in the case of unaided analysis is:

$$R_M^* = \min \varepsilon[r_M(a_M^i)]$$
$$a_M^i \in A_M$$

where $r_M(a_M^i)$ is a distribution of regrets which are expected to result from choosing alternative a_M^i as defined.

ε = expectation operator

In similar function, the minimum "expected regret" when a DSS can be exercised is:

$$R_{DSS}^* = \min \varepsilon[r_{DSS}(a_{DSS}^i)]$$
$$a_{DSS}^i + A_{DSS}$$

If use of a DSS has value in the decision-making process, then we would expect that:

$$R_{DSS}^* \leq R_M^*$$

The value of the DSS in fact should be related to the value of

$(R_M^* - R_{DSS}^*)$. The fact is that positive value can only be expressed by the fact that $R(AM) \leq R(A_{DSS})$ and/or that the variance of $r_M(a_M^i)$ is greater than $r_{DSS}(a_{DSS}^i)$ for one or more i.

The probability distributions of the regret functions with and without the decision support system are likely to be reduced in variance. That is to say, the confidence one would have in a particular regret from a specific alternative a_{DSS}^i would be greater than for the same decision without the DSS (e.g., a_M^i).

The regret associated with a specific alternative a^i is implicitly a function of the resultant state of the "system" given that particular decision, thus we can rewrite $r(a^i)$ for M or DSS as

$$r(a^i) = r(a^i/s^i)$$

where s^i is the resultant system state given a^i.

In more detail we can further derive $r(a^i/s^i)$ from a definition of the regret function. Let $B(s^i/a^i)$ be the benefit of having chosen alternative a^i which resulted in state s^i. The expected benefit of alternative a^i would be

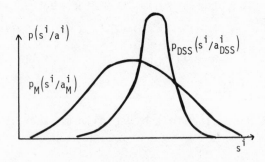

Fig. 1.

$$\overline{B}(a^i) = \int B(s^i/a^i)\, p(s^i/a^i)\, ds^i$$

$$s \in S$$

where $p(s^i/a^i)$ is the probability density of s^i given a^i.

$r(a^i/s^i)$ can be defined as:

$$r(a^i/s^i) = \{\max_{i \in A} B(s^i/a^i)\} - B(s^i/a^i)$$

thus,

$$\mathcal{E}[r(a^i/s^i)] = \int_{s \in S} r(a^i/s^i)\, p(s^i/a^i)\, ds^i$$

$$= \int_{s^i \in S^i \in A} \{\max B(s^i/a^i)\}\, p(s^i/a^i)\, ds^i - \overline{B}(a^i)$$

There is a maximum benefit function associated with each particular level of state s^i; we can define

$$T(s^i) = \max_{i \in A} B(s^i/a^i)$$

thus, the above equation becomes:

$$\mathcal{E}[r(a^i/s^i)] = \int_{s^i \in S} T(s^i)\, p(s^i/a^i)\, ds^i - \overline{B}(a^i)$$

thus,

$$R^* = \min_{i \in A \; s^i \in S} [\int T(s^i)\, p(s^i/a^i)\, ds^i - \overline{B}(a^i)]$$

The above equation indicates that our regret depends upon the difference between the most optimistic benefit to state i over its range S and what our expected response to state i would be. Thus our "regret" depends upon how well we can project our best strategy and how well we can perform on average. This implies that R^*_{DSS} may not be $\leq R^*_M$ if use of the DSS widens our range of potential responses to the point that has a wider potential gap between what actions we conceived of taking versus that action we really did take. Thus, if we define

$$T^*(a^i) = \int_{s^i \in S} T(s^i) \, p(s^i/a^i) \, ds^i$$

then

$$R^*_M - R^*_{DSS} = \min_{i \in A_M} [T^*_M(a^i) - \bar{B}_M(a^i)] - \min_{i \in A_{DSS}} [T^*_{DSS}(a^i) - \bar{B}_{DSS(ai)}]$$

If $T^*_M(a^i) = T^*_{DSS}(a^i)$ for all i then the DSS has not expanded our knowledge of the decision space but has only potentially reduced the variance of our state prediction, and thus $R^*_M \geq R^*_{DSS}$.

CONTAMINANT CONTROL EXAMPLE

An example of DSS benefits is a groundwater contamination study. In on-going work for an agency of the U.S. Government, we have looked at the process of estimating the contamination of ground-water at a government property boundary and our ability to model the effects of containment measures on the migration of contaminant. One of the chemicals in question called DCPD (dicyclopentadiene) has been projected to move over the boundary in different amounts depending upon two containment alternatives: a^1 = Scenario I containment (1500' barrier), and a^2 = Scenario II (4200' barrier plus area capping). Table 1 shows the kilograms of contaminant that are projected to move across the boundary. Table 2 shows the expected costs of the two alternatives.

Table 3 indicates the expected variability of the model concentration predictions for each concentration modal point. The actual "benefit" function (e.g., kilogram across the north boundary) has an intrinsic error bound which is reduced by a factor of $1/\sqrt{n}$ where n is the number of concentration points used to estimate the total contaminant flow. Thus, a ± 184 ppb* bound (out of 500 ppb standard) would result in a $\pm 16\%$ error on predicted kilograms because the predicted kilogram flow of contaminant across the boundary is averaged over $n = 5$ concentration point estimates.

Without the DSS we would not know the Table 1 estimates to any better than $\pm 100\%$. In the case of cost, our estimates would have been only as good as much as $\pm 200\%$. For this reason we could have easily chosen either alternative a^1 = Scenario I or a^2 = Scenario II because we would have known how effective the containment action would be. We would have probably leaned toward Scenario I because of its much lower cost (no surface capping), and we would

---

*ppb = parts per billion concentration.

Table 1. Kilograms Moving Across Boundary in the Example

	Kilograms Over Standard		Total Kilograms	
Year	Scenario I	Scenario II	Scenario I	Scenario II
1980	0	0	3	3
1985	25	0	150	25
1990	105	10	285	40
1995	180	25	350	65

Table 2. Total Example Costs

	Uninflated Dollars (15 Year Operation)
Scenario I	Million $
Boundary Slurry Wall	0.5
Wells	0.4
Decontamination at Boundary	1.0
Total	$1.9
Scenario II	
Boundary Slurry Ext.	2.0
Wells	1.2
Capping Basin A	19.0
Capping Basin F	16.0
Decontamination at Boundary	3.0
Total	$41.2

Table 3. Ground Water Model Predictability*
(Direct Measurement of Observed Test Wells)

DIMP (Standard = 500 PPB)

Standard Error Estimates

	1977	1978
	112	105
90% Confidence Bounds	±184ppb	±173ppb

*Calibrated over years 1974, 1975, 1976.

not have known the extent of our regret in terms of the extent that the contaminant discharged over the boundary.

REFERENCES

Keen, P. F., and Scott-Morton, M. S., 1978, "Decision Support Systems: an Organizational Perspective," Addison-Wesley, Reading, Mass.
Nickerson, R. C., and Boyd, D. W., 1980, The use and value of models in decision analysis, Operations Res., 28(1):139-155, January-February.

APPLICATION OF RISK AND UNCERTAINTY ANALYSIS IN THE PRINCIPLES,

STANDARDS, AND PROCEDURES OF THE U.S. WATER RESOURCES COUNCIL

David C. Campbell

U.S. Water Resources Council
Washington, D.C.

The Water Resources Council, as part of its effort to develop
a set of consistent procedures for water resources planning, has
published guidance for the evaluation of risk and uncertainty in
the formulation and evaluation of management and development plans.
The Council is now coordinating the development of training materials and programs, with the affected agencies, to assist in the
implementation of the procedures. This morning, I will describe
the steps the WRC has taken to date to develop principles, standards, and procedures for the analysis of risk and uncertainty,
and the theoretical and implementation problems that still need
to be solved.

On July 12, 1978, the President directed the Water Resources
Council to "carry out a thorough evaluation of current agency practices for making benefit and cost calculations" and "publish a
planning manual that will ensure that benefits and costs are estimated using the best current techniques, and calculated accurately,
consistently and in compliance with the Principles and Standards
and other applicable economic requirements."

One of the items addressed in the President's directive was
the "uncertainty and risk of costs and benefits" associated with
Federal water resources planning. Based on the work plan dated
October, 1978, which led to the development of the draft "Manual of
Procedures for Evaluating Benefits and Costs of Federal Water Re-
sources Projects," the Water Resources Council proposed a Scope of
Study with two objectives:

1. To develop a procedure for determining those parts of

the economic benefit analysis where risk and uncertainty
are most likely to have a significant impact, and

2.   To develop a procedure for the application of sensitivity
     analysis to economic evaluation of project benefits.

The February 8, 1979, draft of the WRC Manual of Procedures
included a chapter on risk and uncertainty analysis which recommended the use of the "Monte Carlo" simulation technique based on
an estimated triangular distribution of net present value. Major
concerns were brought out about the use of one statistical procedure to measure risk and uncertainty.

The difficulty with this approach is in developing the probability distributions necessary for use in the simulation model.
In most simulation studies, probability distributions are developed
from historical data. However, many times, the accumulation of
historical data is difficult, if not impossible for a water resources project. As a result of this dilemma, the probability
distributions were recommended to be built from subjective probabilities.

There are numerous advantages and some limitations associated
with the application of Monte Carlo simulation to the analysis of
uncertainty in public water resources projects. Some of the more
obvious advantages may be summarizes as follows:

1.   The Monte Carlo simulation technique is a simplified
     approach and may be easily understood by all personnel
     participating in the project formulation, especially
     those making the subjective estimates of benefits and
     costs.

2.   The project planning manager and engineer are generally
     comfortable supplying the estimates for this approach.
     This allows the manager to not only understand the prob-
     lems of risk and uncertainty but also to take an active
     part in its reduction. As a result, the manager is more
     willing to use the model and to accept its results as an
     aid to decisionmaking.

3.   The evaluating agency generally has the expertise and
     resources within their organization to develop the com-
     puterized simulation model. There are numerous routines
     and programs already developed in the area of simulation
     to draw from and to aid in the development of a specific
     model which fits the exact needs of the agency.

4.   The use of the triangular distribution and subjective

estimates offers an opportunity to obtain probabilistic
information when historical data are not available as in
the case of water resources projects.

The proposed simulation technique also has several limitations,
the most prominent of which are:

1. Many agency directors, project managers and politicians
    may be hesitant to accept probabilistic information of
    any type for decisionmaking purposes because such infor-
    mation may imply weakness in the analysis. An uncomfort-
    able feeling with ranges or probabilities may make parti-
    cipation less than complete, especially for those unsure
    of their data or estimating abilities.

2. The project manager may make biased estimates. The mana-
    ger is sometimes hesitant to admit that he is not certain
    about things which he feels he should be extremely cer-
    tain, i.e., the elements of his job. As a result, his
    estimates may be less than candid and yield a less accur-
    ate expected value and standard deviation.

3. The triangular distribution method as recommended in the
    consultant's report does not correlate the components to
    one another, i.e., the relationship of power to irriga-
    tion water supplies in the proposed project.

4. Specific objections to the February draft were (a) the
    use of statistical analysis on one combined parameter
    (net present value) rather than on the individual vari-
    ables, (b) insufficient emphasis on sensitivity analysis
    to isolate the key variables, (c) the need for a step-
    by-step approach rather than jumping immediately to a
    Monte Carlo technique, (d) the emphasis on mathematical
    techniques would make field planners hesitant to use the
    approach.

Because of the comments on the February 8, 1979, draft and the
reluctance to mandate only one out of several possible approaches,
the final rules accepted by the Council describing the procedures
for dealing with uncertainty were general rather than specific.
These were published as Sections 7.13.31-.41 in Subpart B (18 CFR
713) of Part IX of the Federal Register on December 14, 1979,
with an effective date 30 days later.

These procedures, when presented as proposed rules in the
Federal Register on May 14, 1979, also were criticized. The principle thrust of many of the commenters was that while the section
provided the basic principles of risk and uncertainty in water

resources planning, it failed to provide a useable procedure. The procedure as written was judged too general, permissive rather than directive, and lacking in instructive detail. Commenters rated the omission of accepted mathematical and probabilistic techniques. Two commenters stated that the section should be completely rewritten to incorporate current mathematical simulation techniques.

The Council felt that the rules provided valuable, yet flexible, guidelines clearly establishing the need to account for risk and uncertainty in the evaluation of water resources plans and decided to publish the May 24 proposal as final.

In an attempt to remedy the deficiencies and omissions in procedural detail, the Council staff was directed to develop a systematic approach to risk and uncertainty, building on the general approach as published on December 14, 1979. The Council staff, in consultation with the affected agencies, reviewed these procedures and considered several options for strengthening the current rule.

At its March 1980 meeting the Water Resources Council of Members decided to propose no changes. The primary reason for abandoning the search for more specific techniques was because concern was expressed that field planners did not possess the necessary expertise to use sophisticated mathematical techniques. The Council motion authorized its Director to "coordinate the development of appropriate training programs to be used by affected agencies for their field personnel."

An interagency task force has been formed which has outlined their tasks as:

1. The development of an approach to creating an awareness among the participants in the decisionmaking process of the uncertainties of benefits and costs and the necessity to deal with them in a systematic manner.

2. The selection of specific practical techniques for dealing with uncertainty that can be learned in a reasonably short time.

3. The organization of a structure and a process to impart these techniques to the field planners.

There are several reasons given as to why Federal agencies have not adopted improved techniques for dealing with uncertainty.

1. There are no incentives for field planners to analyze variables which are uncertain or risky. In fact, the reward system may encourage the display of "certain" point estimates.

2.   There is little expressed awareness of the uncertainty
     inherent in the estimates of the relevant variables.

3.   There is no consensus at the national level regarding
     techniques for dealing with risk and uncertainty.

4.   There is insufficient expertise at the field level for
     dealing with risk and uncertainty.

5.   It is felt that the evaluation of uncertainty complicates
     and lengthens the planning process.

6.   There may be a reluctance to adopt procedures that might
     forfeit the perceived autonomy of the field planning teams
     in the formulation and evaluation of management and devel-
     opment plans. One commenter argued, "I don't think deci-
     sionmakers (Congress, OMB, Department Heads, COE, etc.)
     can handle many numbers; they rely on planners' judgement."

These obstacles can be reduced by (a) increasing the awareness at all levels of the planning process of the necessity for
and the usefulness of risk and uncertainty analysis, and (b) providing practical techniques for dealing with risk and uncertainty.

I am suggesting a two-level approach to the perceptions of the
problems. We are developing a format for a half-day awareness session on risk and uncertainty analysis. These sessions will be
directed toward a broad spectrum of agency managers and planners
and will attempt to sell an in-depth training program for field
personnel.

The specific details of the proposed training program have
yet to be agreed upon. General topics may include: sources of
uncertainty, dealing with bias, probability distributions, expected
value and other statistical measures, sensitivity analysis, Monte
Carlo simulation models, graphical displays, etc. Many of you in
the audience this morning have knowledge of the theory of dealing
with uncertainty and experience in the application of statistical
techniques to specific estimation problems. We would like to have
your assistance and suggestions as we prepare the training program
for dealing with uncertainty in water resources planning.

RISK ANALYSES APPLICABLE TO WATER RESOURCES PROGRAM AND

PROJECT PLANNING AND EVALUATION*

Ronald M. North

Institute of Natural Resources
The University of Georgia
Athens, Georgia

Water resources project planning and evaluation involve engineering and economic types of analyses; and both are valid only to
the extent that future events are correctly estimated. Each assumption, each conclusion is dependent on the analysts' ability to estimate future events. Regardless of the analysts' experience or the
magnitude of the data base, all estimates of future events are
inherently uncertain.

McKean [1958] suggested the classification of uncertainty for
future events in water resources projects into five groups:
(1) about national objectives, goals and policies, (2) about systems constraints such as resource scarcities, (3) about public
response to proposals, (4) about technological change and
(5) about random events and chance occurrences of recurring events
such as flood and drought both spatially and temporally. The un-
precedented obsession with perfecting the calculus of the benefit-
cost analysis during the decades of the 1930s, 40s, 50s, and 60s
obscured the greatest sources of potential errors (i.e., the uncertainties of future events) by refusing to even recognize the possibility that outcomes were, in fact, mere estimates. The result was

_____

*This presentation is based on previous work done in collaboration
with Dr. Bernard W. Taylor of the Department of Business Administration, Virginia Polytechnic Institute and State University,
Blacksburg, Va., and Dr. K. Roscoe Davis of the College of Business
Administration, The University of Georgia, Athens, Ga. [see reference at the end of this paper to the report by Taylor, Davis, and
North, 1979].

the unquestioned adoption of the benefit/cost ratio as a single-
valued, deterministic, completely certain statement of fact about
economic events (and sometimes hydrologic events) expected 100
years in the future! Even though Eckstein [1961] suggested that
"A more satisfactory approach is to recognize explicitly that project effects should not be predicted as single fixed values but
rather as variables having some probability distribution of possible values" the official guidelines for water resources planning
are extremely timid on the subject. The best suggestions for handling risk and uncertainty are crude beyond belief in this age of
sophisticated systems models and massive data bases. The best
advice to water resources planners for handling risk are to use
(1) a contingency method (judgmentally reduce benefits or increase
costs), (2) limit the planning horizon or (3) add a risk factor
to the discount rate (an impossible option since adoption of formula
discount rates).

Our purpose in this work, which began in 1973, has been to
develop a workable system of risk and uncertainty analysis which
is compatible with existing engineering and economic analyses of
projects, which is easily used at field levels, and which is
easily explained at policy and decision making levels or at public
information events. We believe that consistent results can be
obtained for risk analyses applied to either basic data parameters
or to traditional project functional purposes using a well known
(and frequently used in business) triangular distribution to obtain
a mean and standard deviation for each economic variable in the
project analysis with the diverse data bases available. Although
numerous kinds of distributions are available--normal, binomial,
poisson, beta, weibull, etc.--the triangular distribution will be
the most effective and universally applicable to all kinds of water
resources projects and programs at either the parameter, purpose
of inter-project level of analysis. We would like to demonstrate
the triangular distribution for a multiple-purpose project, working
at the functional purpose level with the planning agency.

EXISTING AGENCY PROCEDURES FOR RISK AND UNCERTAINTY ANALYSES[1]

Existing agency procedures suggest that uncertainty should be
measured yet they offer no methodology to do this. However, all

---

[1]These observations are based on the following agency guidelines:
Corps of Engineers - ER 1105-2-250 and ER 1105-2-921, both dated
10 Nov., 1975; Bureau of Reclamation - Part 112.1.7D and Part
112.1.7G. Plan Formulation dated 12/16/77 and Part 116.2.1C and D;
Soil Conservation Service - "Economics Guide for Watershed Protection and Flood Prevention," Chapter 1, dated 3/1/64.

of these procedures (as related to benefit and cost estimates) are rather superficial and typically do not measure uncertainty.

The Corps of Engineers employs a relative frequency approach for flood control. They develop an expected value based on probabilities of occurrence. However, an expected value is still a single value estimate and does not include a direct measure of uncertainty. In other words, an expected value does not describe a "distribution of values." The corps procedures suggest (for other parameters) that a vigorous statistical analysis to establish certainty is not required but that uncertainty ranges of 0-10%, 10-50%, and over 50% should be designated. This does not address the problem of dispersion of probable outcomes and is uncontrollable.

The Bureau of Reclamation suggests the arbitrary (and uncontrollable) procedures of deducting uncertainty allowances from benefits, adding to costs of adding for contingencies.

The Soil Conservation Service is essentially the same as the Bureau of Reclamation, i.e., adding allowances to benefits and costs as contingency allowances. However, the SCS adds the options of shortening the project economic or physical life or using higher discount rates. These added options for SCS are also uncontrollable and very likely inconsistent with other established practices and regulations (laws) for project life estimates and discount rates.

The Principles and Standards [U.S. Government, 1973] merely define the terms--risk, uncertainty and sensitivity analyses--and suggest that ". . . Net returns of a project should exclude all predictable risk . . ." by deducting from benefits or adding to costs. The Principles and Standards further suggest that allowances for uncertainties must be based on judgment, discussed in project planning reports and considered in project designs. However, no specific procedures are suggested for dealing with risk and uncertainty nor for controlling the risk and uncertainty measurements. The result is that project risks and uncertainties are ignored in project planning.

ALTERNATIVES TO CURRENT PRACTICES

The basic alternative to the methods of risk and uncertainty analyses currently practiced by federal agencies is to develop a probability distribution of benefits and costs which is defined by an average of "mean" value and a measure of variability or "variance." Such a distribution would describe statistically all possible "adverse" and "beneficial" outcomes and their limits. Given such a distribution and its accompanying parameters (i.e., a mean and variance), individual agencies would no longer develop just expected outcomes but also probabilities that this value and alternative values would occur.

The easiest and most direct approach for the development of a probability distribution is to employ historical data. Such data have been used for years to develop probabilities of floods used in the analysis of flood control benefits. However, historical data are not always as available for other benefit and/or cost categories. For example, many items in the categories of recreational, fish and wildlife, navigation, water supply, and environmental benefits and costs have limited data bases that cannot be applied across numerous projects. Each project tends to be unique and independent of other projects, thus, historical data are not easily transferable.

Since historical data for similar projects have limited use in the development of probability distributions it must be complemented with prior judgment about the project being planned. This judgment analysis takes the form of "subjective probabilities" arrived at by agency expertise. This expertise not only consists of judgmental perceptions about the outcomes of the future, but also, reflects experience based on past occurrences (including historical data).

A framework for developing probability distributions from subjectively based probability estimates can be drawn from the private sector. Hertz [1964], in his discussion of the uncertainty involved in private capital investment, presents the concept that probability distributions can be based on "probabilistic" judgments made by the executive decision-maker. In this approach, management takes the various levels of possible cash flows resulting from the proposed investment and makes a subjective estimate of the probability of each of the potential outcomes. In this manner a subjective probability distribution is developed from which a mean and variance are computed for comparison with alternative investments similarly evaluated.

This same framework can be employed in public investment analysis, specifically in water resource project benefit-cost analysis. However, the means by which these probability distributions are developed can differ. The remainder of this paper will be concerned with a description of these alternatives and the development of a preferred approach for reflecting project benefit and cost uncertainty. These alternatives include the triangular distribution, an adjusted triangular distribution, the Weibull distribution, the beta distribution, and the normal distribution. The triangular distribution and the beta, both require minimum amounts of data input (i.e., a minimum of subjective estimation). The other alternatives assume a greater amount of information to describe the distributions. As such, these distributions tend to be more well-defined and, thus, more accurate. (However, one must keep in mind that more information must be available.)

The Triangular Distribution

Since a water resource project is a unique, singular project, the accumulation of historical data is difficult, if not impossible (except perhaps for flood control). As a result of this dilemma, the probability distributions must be developed from a combination of available data and subjective probabilities. The proposed method for developing the probability distributions for the simulation model is similar to the method employed in PERT analysis.

The similarity lies in the method of gathering estimates to be used in probability distributions for the unique project under analysis. In general, three estimates for each project benefit and cost are gathered: a most likely, an optimistic and a pessimistic. Generally speaking, the optimistic estimate is the maximum benefit value (or minimum cost value) that can result from a particular benefit type (i.e., flood control, recreation, etc.), which can be obtained only if unusual good luck is experienced and everything "goes right." The pessimistic estimate is the minimum benefit value (or maximum cost value) that will result only if unusually bad luck is experienced. The most likely estimate is the normal benefit or cost value which should be expected to occur most often if the activity could be repeated a number of times under similar circumstances. This would typically be the single-valued estimate computed under conditions of assumed certainty employed in present evaluation methods.

These three subjective estimates define a triangular distribution which can be used in a simulation model. Since symmetry is not required for this distribution, it is very flexible and it has been shown to be as statistically efficient (and appropriate for simulation models) as other distributions (i.e., the beta).

The triangular distribution described by the three estimates of a particular project benefit yield a probability density function, "$f_X(X)$" (where "X" is the random variate) from the triangular variate "x" (i.e., project benefits). The probability density function is described mathematically as:

$$f_X(X) = \begin{cases} \dfrac{2(X-a)}{(b-a)(c-a)}, & a \le X \le b \\[2ex] \dfrac{2(c-X)}{(c-b)(c-a)}, & b \le X \le c \end{cases} \qquad 0 \text{ , elsewhere.}$$

The cumulative probability distribution function, "$F_X(X)$,"
is calculated by integration of the probability density function,

$$F_X(X) = {}_a{}^c f_X(X)d \ .$$

The Adjusted Triangular Distribution

This distribution is also triangular but requires additional
probabilistic information over and above the three estimates
required in the regular triangular distribution. In this method
[as proposed by Swirles and Lusztig, 1968] a range of possible
values would be estimated for each benefit-cost category. These
probabilities would be assigned to some level of the variable.
For example, a 90 percent chance would be given for a certain
value, a 50 percent chance for another and a 10 percent chance for
another value.

Naturally the values in each section of the triangle are
weights totalling 100 percent. This logically represents a more
well-defined distribution than either the regular triangular or
the beta. However, a distinct problem exists in assigning probabilities to different values, especially for the more subjective
data. It is often quite difficult to make "three estimates" much
less assign probabilities. It is also often difficult to explain
the estimation process for such a distribution to field personnel.

The Weibull Distribution

The Weibull probability distribution is a distribution which
incorporates subjective estimates; but, like the adjusted triangular, it requires substantial information. The Weibull requires
five subjective estimates (rather than three) as follows: the
most likely, the estimated low value, the probability that the
actual value might be lower than the estimated low value, the estimated high value, and the probability that the actual value might
be higher than the estimated high value [Mercer and Morgan, 1978].

These additional estimates naturally provide a more well-
defined distribution (as is the case any time additional information is available). However, the estimation of probabilities of
"how much actual values will differ from estimated values" can be
exceedingly difficult. For this reason, the applicability of this
distribution is viewed with some skepticism, although, if the estimator is confident of his ability and available information, then
the Weibull can be employed.

The Beta Distribution

The Beta distribution was traditionally used in PERT (Project Evaluation and Review Technique) analysis for the same type of estimation process described in the simulation approach which follows this part. However, in the case of PERT, time estimates were made instead of benefit and cost estimates.

Moder and Phillips [1964] note that:

> to estimate the mean, a functional form for the unknown probability curve is required. A likely candidate is the well known beta distribution which has the desirable properties of being contained entirely inside a finite interval, and can be symmetric or skewed depending on the location of the mode relative to the end points. Lacking an empirical basis for choosing a specific distribution, the beta distribution was historically accepted . . .

However, subsequent research has, in fact, shown that there is no a priori justification of the beta as opposed to a triangular distribution. In fact, a comparison of several distributions, showed the triangular to be equally acceptable [Van Slyke, 1963]. The triangular seems to be more "visually" understandable to field personnel and, thus, more easily describable (which naturally contributes to easing the evaluation process).

The Normal Distribution

The normal distribution represents the end of the spectrum in probability distributions in that it is completely defined (i.e., the mean and standard deviation) by historical data in the form of a frequency distribution. As such, it excluded (for the most part) the use of subjective estimates and would be used only when substantial historical data are available and when all project parameters and conditions are expected to be sufficiently representative of the historical conditions reflected in the data base.

For those cases where an alternative distribution form is used in the general simulation model, the probability density function would have to be used. This primarily requires a programming modification. As such, the general simulation methodology is the same for any distribution. However, since most of the benefit-cost categories in water resource projects require some degree of subjective estimation (perhaps excepting flood control) the normal distribution is of limited usefulness either in the field or in the centers of decision making.

A Feasible Distribution and Methodology

Given the absence of data on which to base a frequency distribution for risk, one must devise some combination of available data and subjective estimates of the reasonable range of likely outcomes and estimate the uncertainty in a controlled procedure. The most feasible approach to this problem in the planning and evaluation of water resource projects would be to adopt the triangular distribution and to adapt a simulation approach to develop the probable outcomes on the basis of three controllable estimates and a definitive procedure to generate sufficient "observations" for a mean and standard deviation and a confidence interval for selected data parameters (population, income, etc.) or preferably for major project components (benefits by purpose, costs, benefit-cost ratio, net present value).

THE SIMULATION APPROACH

One of the most frequently used (and easily applicable) methods available for calculating the mean and variance parameters of a benefit-cost probability distribution is "Monte Carlo" simulation based on an estimated distribution [Hertz, 1964; McKean, 1958]. The term "Monte Carlo" has become almost synonymous with simulation in recent years due to its popular and varied uses in simulation. Monte Carlo is a mathematical technique of selecting numbers from a probability distribution for use in a particular trial run in a simulation study [Meier, 1969].

In order to develop a simulation model using the Monte Carlo method, it is first necessary to determine probability distributions for the various benefit and cost parameters of a water resource project. These distributions will subsequently be employed in the simulation model which will yield numerous variable outcomes. These outcomes will result in a distribution of possible benefit-cost criterion values for the proposed water resource project. This distribution will ultimately yield a mean (average) outcome and a standard deviation as a measure of variability. The generation of these two statistical parameters accomplishes the objective of including uncertainty in the water project economic evaluation process.

A brief example will demonstrate the steps of the simulation process. Consider the development of the probability distribution for one benefit area, navigation. The agency official(s) responsible for deriving the three "subjective" estimates for this benefit would employ all available expertise, in addition to any applicable technical or historical data. (The same process could be used to develop the pessimistic and optimistic estimates as is used to develop the single estimate now made for each benefit category or project.) Let us suppose that the three required estimates were gathered as follows:

a = $30,000 = pessimistic estimate

b = $40,000 = most likely estimate

c = $60,000 = optimistic estimate

These values can be substituted directly in the process genera-
tor for the triangular variate "X"

$$
X = \begin{cases} a + \sqrt{(b-a)\ (c-a)\ U}\ , & 0 \le U \le \dfrac{(b-a)}{(c-a)} \\[3mm] c - \sqrt{(c-b)\ (c-a)\ (1-U)}\ , & \dfrac{(b-a)}{(c-a)} \le U \le 1 \end{cases}
$$

$$
X = \begin{cases} 30{,}000 + \sqrt{(40{,}000 - 30{,}000)\ (60{,}000 - 30{,}000)\ U}, & 0 \le U \le .33 \\[3mm] 60{,}000 - \sqrt{(60{,}000 - 40{,}000)(60{,}000 - 30{,}000)(1-U)}, & .33 \le U \le 1, \end{cases}
$$

Now, a random number, "U," is drawn from a table of random numbers and "U" equals .20. Since .20 is within the range, $0 \le U \le$
.33, it is substituted in the top equation to yield a value for "X"
as follows:

$X = 30{,}000 = \sqrt{(10{,}000)\ (30{,}000)\ (.20)}$

$X = \$30{,}000 + 7{,}746$

$X = \$37{,}746$

This is one (of many) navigational benefit value. The same process would now be repeated for all other benefit values (i.e., flood
control, recreation, etc.) and then summed for total project benefits. The same process is conducted for all cost values (i.e., capital cost, operation and maintenance, etc.). The total project benefits and total costs would then be substituted in the selected
benefit-cost criterion (i.e., benefit-cost ratio and/or net present
value) with the designated discount rate and project life to yield
one benefit-cost value. This represents one simulation run and,
thus, one outcome in the probability distribution. By repeating
this same process for numerous runs, (i.e., as many as 100 to 1,000
runs) many outcomes are developed which form a probability distribution from which a mean and standard deviation are calculated. A
total case analysis employing this process is presented in Taylor and
North [McKean, 1958; Taylor and North, 1975; Taylor and North, 1976;
Taylor, Davis, and North, 1979].

For the example presented above, one navigation benefit value was computed <u>manually</u>. In an actual application, it would be much more efficient to use a computerized simulation program. A program can be developed to perform these operations using two prewritten subroutines "DRAW" and "TABLE" developed by T. J. Schriber [1969]. These subroutines are written in FORTRAN and anyone with average programming skills can develop a program from them to perform the benefit-cost simulation.

Advantages and Limitations of the Simulation Approach

There are numerous advantages and some limitations associated with the application of Monte Carlo simulation to the analysis of uncertainty in public water resource projects. Some of the more obvious advantages may be summarized as follows:

(1) By including the uncertainty associated with the economic analysis of water resource projects, the probability of errors in the estimation of benefits and costs is minimized. That is, the range of all possible benefit and cost values are within the maximum and minimum estimates made by the project planning manager.

(2) The Monte Carlo simulation technique is a simplified approach and may be easily understood by all personnel participating in the project formulation, especially those making the subjective estimates of benefits and costs.

(3) The project planning manager and engineer are generally comfortable supplying the estimates for this approach. This allows the manager to not only understand the problems of risk and uncertainty but to take an active part in its reduction. As a result, the manager is more willing to use the model and to accept its results as an aid to decision-making.

(4) The evaluating agency generally has the expertise and resources within their organization to develop the computerized simulation model. There are numerous routines and programs already developed in the area of simulation to draw from and to aid in the development of a specific model which fits the exact needs of the agency.

(5) The use of the triangular distribution and subjective estimates offers an opportunity to obtain probabilistic information when historical data are not available as in the case of water resource projects.

The proposed simulation technique also has several limitations, the most prominent of which are:

(1) Many agency directors, project managers and politicians may be hesitant to accept probabilistic information of any type for decision-making purposes because such information may imply weakness in the analysis. A feeling of uncomfortableness with ranges or probabilities may make participation less than complete, especially for those unsure of their data or estimating abilities.

(2) The project manager may make biased estimates. The manager is sometimes hesitant to admit that he is not certain about things which he feels he should be extremely certain, i.e., the elements of his job. As a result, his estimates may be less than candid and yield a less accurate expected value and standard deviation.

Application of the Simulation Results

Once the uncertainty inherent in a benefit/cost criterion value has been developed in the form of a mean and standard deviation, it can be employed in several ways.

One possible evaluation method is the "coefficient of variation, V," probability of a benefit-cost ratio being less than 1.0. If this probability proved to be too high (as perceived by the evaluating agency), then a re-evaluation of the project might be deemed necessary.

These simple statistical methods represent a few of the more obvious vehicles for applying uncertainty (as reflected by the mean and standard deviation) to benefit-cost analysis.

REFERENCES

Eckstein, Otto, 1961, A survey of the theory of public expenditure criteria, in: "Public Finances: Needs, Sources and Utilization," National Bureau of Economic Research, Princeton University Press, Princeton.

Hertz, David B., 1964, Risk analysis in capital investment, Harvard Bus. Rev. 42:95-106.

McKean, Ronald N., 1958, "Efficiency in Government Through System Analysis," Wiley, New York.

Meier, Robert C., Newell, William T., Pazer, Harold L., 1969, "Simulation in Business and Economics," Prentice-Hall, Englewood Cliffs, N.J.

Mercer, Lloyd J., and Morgan, W. D., 1978, Measurement of economic uncertainty in public water resource development: an extension, Am. J. Ag. Econ., 60:241-244.

Moder, Joseph J., and Phillips, Cecil R., 1964, "Project Management with CPM and PERT," Reinhold, New York.

Schriber, T. J., 1969, Simulation of probabilistic cash flows and
        internal rates of return, in: "FORTRAN Case Studies for Business
        Applications," Wiley, New York.

Swirles, John, and Lusztig, Peter A., 1968, Capital expenditure
        decision under uncertainty, Cost Management, 42:13-19.

Taylor, Bernard W., and North, Ronald M., 1975, "A Simulation
        Approach to the Analysis of Uncertainty in Public Water Resource
        Projects," USDI/OWRT Project No. A-052, Ga., Athen   Georgia:
        Institute of Natural Resources, University of Georgia, and
        Atlanta, Georgia: Environmental Resources Center, Georgia
        Institute of Technology.

Taylor, Bernard W., and North, Ronald M., 1976, The measurement of
        economic uncertainty in public water resource development, Am.
        J. Ag. Econ., 58:636-643.

Taylor, Bernard W., Davis, K. Roscoe, and North, Ronald M., 1979,
        "Risk, Uncertainty, and Sensitivity Analyses for Federal Water
        Resources Project Evaluation," U.S. Water Resources Council,
        Washington, D.C.

U.S. Government, Water and Related Land Resources, Establishment of
        Principles and Standards for Planning, 1973, Federal Register,
        38, No. 174, Part 3, Washington, D.C.

Van Slyke, Richard M., 1963, Monte Carlo methods and the PERT prob-
        lems, Operations Res., 11:839-860.

RISK ASSESSMENT FROM A CONGRESSIONAL PERSPECTIVE

James W. Spensley

U.S. House of Representatives
Washington, D.C.

Congress is becoming increasingly aware of the use of new tools such as risk assessment[1] to assist them in their public policy making function. The growing complexities of our technologies and their possible impacts to our economic, environmental and social systems have made the task of formulating public policy with respect to their use nearly impossible. Thus, as government has become more concerned with technology, technology has become more political. Because the process of formulating public policy on the important issues is dependent on the comprehensibility of relevant information, the collection, analysis and presentation of that information is essential in order that intelligent options can be evaluated and the appropriate courses of action chosen.

The use of risk assessments in the policy formulation and decision-making process is one of several tools which is receiving more attention. During the first session of the 96th Congress, two Congressional committees in conjunction with the American Association for the Advancement of Science held a two-day forum on risk/benefit analysis in congressional science and technology policy decisions.[2] This forum brought together a unique mix of experts and politicians to consider the application of risk assessments in the congressional arena. This forum helped draw attention to the use of this new tool in dealing with the technological and policy choices facing the Congress. Although the forum participants did not draw a consensus on how risk assessments should be used in policy decisions, the explorations of the issues surrounding its use helped delineate the boundaries of its utility.

It is generally recognized that risk is present in all activities and that the public has a very poor understanding of risk.

Though every Member of Congress performs his/her own mental risk
assessment on important political issues, a general recognition and
standardization of that process has been the subject of considerable discussion. Clearly, some Members would like a simple formula
for evaluating the risks, costs and benefits of any proposed action
while others recognize the vulnerability of such an approach. The
only widespread agreement seems to be that risk assessments are
difficult to perform and should be used with caution.

In that regard, it is interesting to note that two very different roles for Congress have been suggested. One opinion is that
Congress generally should not be involved in day-to-day decisions
regarding the regulation of risk, but should develop policies and
risks acceptance criteria which can be applied by the appropriate
regulatory agency. It is argued that risk assessment is not appropriate for congressional considerations because political pressures
upon Congressmen not to appear reckless with human lives would lead
to an over-emphasis on safety. The contrasting opinion is that
risk assessment decisions are too subjective to be left to experts
who have their own biases and so must be performed by the Congress.
After all, the Congress best reflects the values of their constituents who elected them to represent them on such important policy
matters.

The most recent Congressional action concerning risk assessments was the hearings held by the Committee on Science and Technology on two proposed bills.[3] The two bills represented differing
approaches to the utilization of risk assessment in public policy
formulation. One approach introduced by Congressman Don Ritter
of Pennsylvania[4] calls for the establishment of a federal mechanism
to apply and promote the understanding and appreciation of comparative risks in scientific, technological and related matters and to
assist federal, state and local governments, private industry and
the public in making intelligent comparisons and evaluation of
those risks. His proposed legislation would give responsibility
to the Office of Science and Technology Policy (OSTP) within the
Executive Office of the President to review the mechanisms which
the federal government now uses for comparing risks and to the
maximum extent, develop and implement procedures to be used in
making such comparisons in the future. Congressman Ritter's proposal also requires the OSTP to develop a plan for educating the
public and the government in the use of risk assessment. Moreover, it sets forth some minimum guidelines to be considered in
the use of risk assessments in the decision-making process. Overall, this approach of exhortation and education was considered
generally acceptable by both the Committee and the Administration.

The other approach is represented by a bill introduced by
Congressman William Wampler of Virginia.[5] This bill calls for the

establishment of a National Science Council to decide questions of scientific fact which arise in agency adjudications involving the regulatory process. Congressman Wampler's bill would establish this new Council in the Office of Science and Technology Policy for the purpose of determining scientific fact which might arise in an agency adjudication to determine the harm of any substance to human health. The Consumer Product Safety Commission, The Secretary of Health and Human Services, the Administrator of the Environmental Protection Agency, the Secretary of Agriculture, the Director of the Occupational Safety and Health Commission, and others are specifically directed to submit all questions of scientific fact in dispute to the Council for determination. Their findings must be used by the agency in their decision-making process. The inherent presumption is that once scientific facts have been established, the processes of regulation and execution of the laws of the Congress would be made easier and without the delay and controversy that exists today. Clearly, this presumption is the subject of considerable discussion.

This latter legislative proposal was also the subject of hearings by the House Committee on Agriculture[6] concerning the issue of nitrates. Nearly half of the testimony concluded that if such a Council existed for purposes of determining the scientific facts involved that the nitrate controversy would not exist.

Although neither of the Committees which held hearings on these proposals found it prudent to take further legislative action during the 96th Congress, it is likely that both proposals will again be considered in the new Congress. Both proposals resulted, in part, from a frustration with the difficulty of making hard choices on complex issues where a multitude of values exist.

As an interested observer in the use of risk assessment in public policy making, I have concluded that the political nature of the Congressional process makes the use of risk assessments particularly difficult. There is a frequent tendency to handle tough political problems by grasping handy "crutches" to simplify the decision-making process. Because of the complexity of the issues and problems presented, Congress has a penchant to deal with the means rather than the "ends." In other words, the ultimate congressional solutions are embodied in process and vague criteria rather than explicit statements of the objectives to be attained. In part, the rationale for this approach is that the agencies and others have greater expertise to consider the specifics of any case and wider discretion to administer to the anomalies than does Congress. Thus, the use of risk assessment does not carry the same weight in fashioning those "means" as it would if genuine consideration of the options of "ends" were the focus of attention.

The Congress lacks a familiarity with the use of risk assessments. The Members of Congress are lay persons and tend to communicate in lay terms. They are unfamiliar with the use of risk assessment as a tool in sorting out difficult issues. As a result, the use of risk assessments is not evident in the consideration of important legislative issues. Moreover, it is not normally expected or requested by the Members in their deliberations. If a risk assessment were presented, there would be a natural mistrust of its meaning, particularly if the conclusions presented are contrary to the Members perceptions. Without the proper understanding of the methodology of risk assessments and its constraints, it is more than likely to be misused.

Another factor which limits the use of risk assessments concerns the politicans' reluctance to talk about "risk." Instead, it is the character of politicians and statesmen alike to talk about "opportunities," not risks. It is better to talk about the advantages and benefits of a particular proposal or technology than to mention, much less emphasize, the cost or risks to the public. To suggest that the favored approach advocated by a particular Member or Committee would kill or maime fewer people than the other approaches is to suggest that perhaps none of the options are worth pursuing. Therefore, the use of risk assessments reflects a "negative approach" to viewing the alternative public policies available.

Moreover, the congressional process lacks the proper framework for utilizing risk assessments in decision-making. Recognizing that risk assessments are only as good as the quality of information used in the process, it is important that quality control be exercised in the collection, consideration and application of that information. In the legislative process, the principle mode for information collection is the congressional hearings. It is in that forum that scientific data and expert opinion is solicited before the Members of Congress. The hearing process is by far the worst method for exercising quality control or in gaining an understanding of the data presented. It does not provide a convenient or appropriate opportunity to examine the assumptions, methodologies or value judgments inherent in the conclusions or recommendations made by witnesses. As a result, the "record" developed remains immune from proper scrutiny and testing. Clearly, it would be unwise to rely on such a record for performing any risk assessment.

A similar deficiency exists in the support services to Congress for performing relevant risk assessments. For the reasons discussed above, congressional staff and support organizations such as the Office of Technology Assessment and others, do not routinely, if ever, prepare risk assessments. The very few that have been done have been the subject of considerable controversy.

Finally, one intractable issue surrounding the use of risk assessment is the exercise of value judgments. In the preparation, execution and presentation of any risk assessment there is a need for the exercise of numerous value judgments. The framing of the question or issue to be considered, the establishment of data parameters, the selection of the methodology, and the presentation of those results all involve the exercise of some value judgments. Then, as a final matter, the application of any risk assessment to the decision making process requires the decision maker to weigh the importance of the risk assessment and its conclusion in the resolution of the matter at hand. To presume that either the methodology or the application of risk assessments can be devoid of such judgments demonstrates a misconception of its character. It is in the ultimate exercise of the decision makers' judgment as to the importance of the risk assessment that belies the notion that the use of risk assessments can avoid the tough political decisions. To argue over whether the Congress or the analysts is better prepared to make such value judgments misses the mark in understanding that it is Congress that will ultimately have to make the final policy decision.

The success of risk assessment as an analytic tool in the congressional public policy making process will depend first on educating the Members as to its proper use as well as its limitations; and second, on increasing the Members familiarity with its use in the decision making process. The future consideration of legislative proposals such as those that have been considered by the 96th Congress will certainly assist in making Members aware of risk assessment as a useful analytic tool.

REFERENCES

1. Risk assessment is intended to include risk/benefit analysis,
       comparative risk assessment and cost/benefit analysis.
2. "Risk/Benefit Analysis in the Legislative Process," Joint
       Hearings before the Subcommittee on Science, Research &
       Technology of the Committee on Science and Technology and
       the Subcommittee on Science, Technology and Space of the
       Committee on Science and Transportation, July 24-25, 1979,
       Print No. 71.
3. Hearings of the Subcommittee on Science Research & Technology
       of the House Committee on Science, Technology, May 14-15,
       1980, on H.R. 4939 and H.R. 6521.
4. H.R. 4939, Introduced by Congressman Ritter, July 24, 1979.
5. H.R. 6521, Introduced by Congressman Wampler, February 13, 1980.
6. Hearings of the House Committee on Agriculture, September 16,
       1980.

RISK ASSESSMENT: THE ROLE OF GOVERNMENT IN A MULTIPLE OBJECTIVE FRAMEWORK

Warren A. Hall

School of Engineering
Colorado State University
Fort Collins, Colorado

> The rain it raineth on the just
>   and also on the unjust fella;
> But chiefly on the just, because
>   the unjust steals the just's umbrella.
>
>   Lord Bowen

One could argue rather persuasively that the only proper role of government is that of risk mitigation in a multiobjective framework. This might require broader or narrower definitions of these terms than their customary usage, particularly that of "proper," but by and large these are precisely the kinds of interpersonal problems which long ago constituted the basis for the original creation of governments. It is admitted that such governments, once created, have commonly usurped broader roles, but the basic reasons for their existence and their character reflect a necessity for a community or group action to avoid or mitigate risks associated with accomplishing the many objectives of the people concerned.

One basic problem has been exemplified by Hardin in what he terms "the tragedy of the commons." In this example, the common pasture shared by the entire community was the resource. As a common pasture, each resident had a right of use for grazing his own animals. Unfortunately, the optimal policy for each person, acting as an individual decision maker, was (and still is) to put as many animals·on the pasture as he could. At some point in time, the affluence, in terms of animals owned by the members, became sufficiently high that the pasture was overgrazed and destroyed as a

production resource. The result was far less benefits for all concerned than could have been provided indefinitely by proper management under community decision.

Another basic problem is that which Mr. Howard Crook (Orange
County Water District) frequently characterized as "people who take
the profit and leave the problem." When individual actions are
taken, they may result both in benefits and in costs to others who
would otherwise not be involved in the decision to take the action.
Economists refer to these effects as "externalities." Actions such
as murder and theft are the most obvious of these, but there are a
great many other actions, including actions taken to develop water
and other resources, with similar, if less vital consequences. In
all cases, the role of government has been to protect those not concerned with the decision against these costs and/or to assure that
all beneficiaries share the costs associated with their benefit.

This role of government is probably the most difficult to
accomplish equitably. Nevertheless, it probably constituted the
original rationale for the creation of the governments of ancient
Egypt and Mesopotamia. It also constitutes the basis for allocation
of certain of the costs of water projects to the general public as
indirect beneficiaries instead of the direct beneficiaries. This is
not to argue that any of these roles are correctly accomplished.
However, they clearly are governmental in nature.

A third basic problem deals with economies of scale or equivalent in which, by a community action, all concerned can accomplish
certain of their objectives with less unfavorable impact on other
objectives (costs). Here the whole is greater than the sum of the
parts. "In unity there is strength." "We must all hang together
or we shall certainly hang separately." There are many such quota-
tions, all of which demonstrate the necessity of a governmental role
of some type, by means of which community decisions can be made and
enforced against the members of that community, if the need should
arise.

Although not entirely obvious, risk mitigation is implicit in
every one of these situations as well as being explicit in some of
them. Presumably, if everyone understood the situation and made his
own decisions to conform to the needs of his society, all of the
actions that a government might make could be made as an aggregate
of individual decisions. However, there is a substantial risk that
some members of the society will not understand the situation or
will attempt to exploit the common interest in favor of his own
benefits.

To prevent such risks from materializing, there must be a mechanism established for a community decision and for enforcement of that

group decision. Such community decision mechanisms are collectively "governments" in the broadest sense. The term would include the government of corporations as well as the more commonly used defini- tion of national, state, provincial, county, and city governments. The only real difference is the scope of authority provided for in the "agreement to agree," in the precedence of authority where con- flicts arise, and to a lesser extent, in the definition of that which shall constitute "agreement."

There are countless examples of this role of government. Fur- thermore, history has shown that when one group of "decision makers" continues to make its decisions without regard to significant adverse impacts on the objectives of the public as a whole, government inter- vention has occurred, bringing the power and authority of a higher level of government in the hierarchy to curtail the scope and type of decisions allowed the original group.

As indicated in a companion paper at this conference, the funda- mental purpose of water resources development is one of meeting a host of objectives of many members of our society whose accomplish- ment is generally limited by inadequate water resources. Further- more, the purpose is not so much to "create more water," but rather to provide a better level of assurance that water will be available when needed, where needed, and in an appropriate quality for the intended uses, thereby allowing us to meet that host of objectives. Alternatively it could be to provide a better level of assurance that it will not be present, where and when it would be deleterious to those objectives.

These requirements argue that some form of government will be essential in order to meet them. In theory, at least, this can be a private corporation type of government as well as public govern- ment. Public government, however, is involved here to an important extent because water resources development not only impinges on the objectives of direct users but also on the objectives of indirect users and on non-users who could otherwise utilize the same resource to accomplish their own spectrum of objectives.

If any persons act to develop a water resource as a private venture, they may acquire the necessary power of decision with which they may accomplish the modification of the hydrologic availability of the natural resource to one more reliable in time and place of use. However, the nature of the physiography associated with the hydrologic cycle is usually such that other individuals or groups could take similar actions to nullify the reliability expected to be achieved. In such cases the reliability is illusory rather than real.

Here the function of government is one of providing security
(reliability) in the form of a system of legal rights of use. While
these take different forms depending on the public governmental unit
concerned and the characteristics of the resource and its potential
uses, it is patently clear that these systems of water rights are
expressions of a public governmental role in risk assessment and its
minimization. In essence the risks of maintaining one's right to a
water supply in time and place were assessed and a scheme for effectively providing for assurance of continued use was designed. In
most cases these schemes include provisions for identification and
prevention of misappropriation as well as formalizing a legal right
of use to individuals as a property right.

There are other aspects, however, which required a more active
role on the part of public governments in water resources development.
Chief among these are (1) the competition which exists among potential users, both present and future, and (2) the externalities, or
impacts, favorable and unfavorable, which extend beyond the immediate
entities involved in the development of a water resource.

One needs only to review the history of virtually any major
water development project to understand the role of public government as the arbiter of competitive claims. For the most part in the
past, it has usually been able to take advantage of the fact that a
single project is capable of serving many purposes and users, with
a scope broader than those of any one person or organized group.
It has also been able to take advantage of economies of scale to
serve more objectives and persons at substantially less cost to
each than would be the case of an individual enterprise.

Not quite so apparent in the history of water projects is the
second aspect requiring a role for government, that of the externalities, or impacts on others. These are certain to occur in any
case, and are extremely important in any major project. Here the
problem to be addressed is that of equity, i.e., equitable sharing
of benefits and costs, economic and noneconomic, whether these fall
on direct users or indirectly on others.

The problem, of course, is the definition of "equity" when, as
in the usual case, these impacts are distributed rather nonuniformly
over the society concerned. Indeed, this must have been a major
force in the original creation of civilization, i.e., civil public
government, as contrasted with purely military or religious gover-
nance of personal and social behavior. The works to be accomplished
by a community, by actions directed toward control of the factors
most visibly affecting the general welfare, were essentially public
works. In order to assure their accomplishment and to assure maintenance of the benefits therefrom, long-lasting agreements were

required on how the benefits were to be shared and how the costs
would be equitably distributed to those who were beneficiaries.

These characteristics of the first public works are still basic
to modern public works. They differ only in degree of complexity,
the scope of the interactive process, and the type of "agreement to
agree" utilized as the basis for authority for action. Conquest as
the basis has largely (but not entirely) given way to political
process as the basis. In the United States, the processes are collectively referred to as "democratic representative government," in
which duly elected representatives of people are given necessary and
sufficient authority for action by the "agreement to agree" (our
federal or state institutions, city and county charters, etc.).

There is yet another role of democratic government in water
resources planning. Despite the modern emphasis on a presumed "will
of the majority" and "one man--one vote," neither constitutes the
basic requirement for successful democracy. Rather, a careful study
of all the "democratic government" experiments throughout history,
including present-day governments, rather emphatically illustrates
the absolute necessity for a de facto obligation on the part of the
"majority" to seek an equitable compromise with the concerns of the
minority in such a manner as to minimize the adverse impacts of the
decisions which, by the authority of constitution or charter of the
government entity, the majority may choose to implement.

History is replete with examples of "democracies" in which
"majority rule" was considered a "right" independent of concerns
of the opposing minority. All have failed or are currently in the
process of failing, often in a monotonously repetitive manner. The
French Revolution of the late nineteenth century is an example, perhaps exaggerated. Lacking a proper understanding of the fundamental
requirements of democracy, many other such experiments have started
nobly, then faltered and collapsed as the concerns of minorities
were crushed under the "will of the majority," the latter being suc-
cessively splintered and crushed to the point where no majority
could be raised to govern. In virtually all of these examples, a
"strong man" in the form of a Napoleon, a Stalin, or a Hitler had
to appear on the scene, usurping the democratic system, replacing
it with an autocratic system.

As the problems, and the system of regulations and public works
intended to resolve them, have become more complex and sophisticated,
it has become necessary for representative democracy to be assisted
in the creation and evaluation of alternative systems for accomplishing public objectives, subject only to final review, amendment,
and/or veto by the proper legislative body. For water-related legislation, this assistance is provided in this country through water
resources planning agencies.

In this sense, all water resources planning which leads to recommendations for enactment by legislative bodies must be conducted as
a surrogate process for the legislative negotiations which would
otherwise need to take place. They are legislative rather than executive responsibilities and authorities under our federal constitution,
as well as under most state constitutions and city and county charters. This is the case even though the organizations which accomplish the actual studies have been placed in the executive branch.
Many of the problems which have developed between the Executive and
the Congress with respect to these and similar matters are the direct
result of a misunderstanding (often mutual) between the persons in
the executive branch and persons in the Congress. It is an efficient
arrangement and in some respects it is also effective. To retain its
effectiveness in a constitutional representative democracy, it must
be understood by all concerned. In particular, the executive agencies
must understand that they have been delegated what is essentially a
legislative responsibility, that of seeking the equitable compromise.

With all of those background comments in mind, the appropriate
role of government in risk-benefit analysis for multiple objective
planning for water resources would include, but not be limited to,
the following:

1. Identify, evaluate, and take necessary community action in
    mitigation of individual risks and community risks, to the
    extent that these limit or militate against actions consid-
    ered to be desirable by the community served by that govern-
    ment.

2. Identify, evaluate, and take necessary community action in
    mitigation of risks (individual and community) to the
    extent that these are caused by individual or group actions
    considered to be undesirable by the community-at-large.

3. Provide institutional and technical assurance that nominal
    or implicitly understood levels of reliability of service
    (including right of service) will be maintained.

4. Provide a mechanism for objective investigation and
    creativity toward the identification and implementation
    of the equitable compromise between short- and near-term
    majority and minority interests, and between these and
    long-term or future interests.

All of these roles of government are implicitly involved with
multiple objectives and associated risks. The latter, in turn,
are nothing more nor less than the aggregate of individual objectives and related risks, modified only by the degree that conditions
which invoke the roles of government exist: i.e, (a) economies of

scale which can make the whole greater than the sum of the parts,
(b) "tragedy of the commons" type situations in which individuals'
optimal decisions result in less for all concerned, and (c) existence of major externalities to individual decisions which would
allow some to benefit at the expense of others without recourse.

To a reasonable extent these roles have been accomplished in
water resources planning in the past. Unfortunately a serious problem is being generated by the almost incomprehensible progress in
mathematical modeling and computer technology.

Only a few years ago, water resources planners were forced to
exercise considerable experienced judgment regarding the factors
involved in government's role. In particular, the scientific or
quantitative models which could be "solved" not only required substantial simplification, but it was very apparent to all concerned
that such simplifications had to reflect well-considered and experienced judgments of all the qualitative matters involved. Furthermore, the mathematical solution itself had to be subjected to further
experienced professional and legislative judgment before a satisfactory problem solution could be designed.

If all progress in mathematical modeling were unformly distributed over the spectrum of factors which were reflected almost completely by judgment in prior years, perhaps there would be no difficulty. This has not been the case. Far more progress has been made
in computer software than in mathematical modeling and far more
progress has been made in mathematical models of certain physical
elements of the problem than in other areas, particularly the quantification of objectives with cardinal numbers--i.e., numbers which
assert that 10 is not only larger than 5 (ordinal measure) but is
precisely two times larger.

As a result of this situation, we can design and solve complex
mathematical models which give an erroneous impression that they
are comprehensive, yet no one really understands them adequately.
By this comment, I do not wish to imply that the designers of such
models do not know or understand the mathematics of their models or
the physical-economic factors they have included therein. What I
am concerned with is an understanding of the models with respect to
what is not included in the mathematical logic, particularly the
implications for those matters which still cannot be properly represented by cardinal measures.

At the risk of some controversy, let me cite an example. Our
mathematical optimization models, in and of themselves, are logically
incapable of optimizing more than one objective expressed as a scalar,
cardinal function of the actions that might be taken. Because mathematical optimization is indeed a very powerful tool for decision
making, this has led to an almost universal assumption that net

economic benefit, expressed in terms of dollars, is in fact the primary objective of all water resources planning and development.

Indeed, economic efficiency has been raised to such a status by
the Principle and Standards, thereby adding credence to the otherwise
blind but convenient assumption that this is the quantity to be maximized. In the process, the host of true objectives, including risk
mitigation, have been relegated to the status of an "account." Only
environmental objectives have survived, and even here, practice suggests an intention that their use be limited to negating otherwise
economically desirable projects. The worst environmental "disasters"
of the nation are given virtually no attention or are exacerbated by
public action.

If we start with the stated intentions of the Congress (by constitution the sole policy-making body for the nation), it is clear
that economic feasibility has been established as a constraint
rather than an objective, ". . . if the economic benefits, to whomsoever they may accrue, exceed costs . . ." is the essence of the
policy established by the Congress as a criterion of feasibility.
It has never stated that the difference, or the ratio, or any other
function of benefits and costs should be maximized.

Clearly this policy is intended to allow consideration of other
objectives as being more important than economic objectives so long
as the criterion of economic feasibility was met.

Indeed, one might argue that, if economic efficiency is the
sole criterion for water resources planning, there should be no
governmental role. Such a criterion, by its very definition,
would appear to exclude all of the proper roles of government except
perhaps that of assuring the property right of use. As indicated by
the language of the Congress, the net of economic benefits and costs,
as a scalar index of performance, is completely indifferent to "whomsoever they may accrue."

Most of us, including myself, have been guilty of this erroneous assumption in the construction of mathematical optimization
models. Sometimes, in the analysis of some component, such as minimization of the direct economic cost of a set of physical facilities
to accomplish a set of other objectives at minimum levels, the arti-
fice is extremely useful even if not strictly correct. However, it
is not and will not be even approximately correct for water resources
development as a whole, unless and until all objectives of all
people (including the distribution of objective achievement) can be
expressed in monetary units using cardinal measure. That is, they
must be capable of being added in the physical-social sense as well
as numerical sense in any desired order and still get the same

correct result. This we cannot do even approximately with the
present state of economic science.

When and if, particularly if, this millenium ever comes to pass,
the role of government in multiple objective water resources planning
and management will be reduced to a simple administration of water
rights institutions.

Until that day, the role of government remains as it always has
been, one of providing a mechanism for community decision making to
minimize the probability of exploitation of the common good by individuals for temporary personal gain, to maximize the probability
that the whole will indeed be greater than the sum of the parts, and
to maximize the probability that the beneficial and nonbeneficial
impacts are distributed as uniformly as possible.

The task ahead is not at all to develop more analytical models
to maximize the net economic benefits "to whomsoever they may
accrue." This is in reality only policy constraint (incidentally
applied only to a very small number of federal investments). Rather
it is to learn how to elaborate the nebulous social well-being objectives and to present the corresponding impacts of water planning
and management on these primary objectives to the people and to the
policymaking branch of government in as factual a manner as may be
possible.

This is the proper role and the only role of public government
in a multiobjective framework. As institutional surrogates for
government, it is also the proper role of water resources planning
and management in a democratic society.

IMPLEMENTING THE RISK AND UNCERTAINTY PROVISIONS OF THE

PRINCIPLES AND STANDARDS

David C. N. Robb

Office of Comprehensive Planning
Saint Lawrence Seaway Development Corporation
Washington, D.C.

The new risk and uncertainty provisions of the Principles, Standards and Procedures undoubtedly stem from a sincere desire to provide decision makers, reviewers, and interested parties with more and better information. There is also clearly an intent to ensure explicit consideration and documentation of the risks and uncertainties inherent in the individual and aggregrate benefits, costs, and environmental and other societal impacts which might result from the recommended plan and from the range of alternatives considered in arriving at the recommended plan. These are laudable aims, with which I fully concur. However, implicit in the requirement for inclusion of risk/benefit analyses in water resources planning documents are increased planning costs in terms of both time and money and, presumably, increased benefits in terms of better, more reliable (less risk-inherent) projects. Thus, one might ask whether the benefit/cost ratio for requiring risk/benefit analysis will be greater than 1.0. Implementation of the requirement implies an affirmative answer, but the question has not been explicitly resolved.

Economists, statisticians, and operations researchers should not be surprised to learn that many agency people view risk and uncertainty analyses not only as tools for doing a better planning job, but also as weapons with the potential for seriously or even fatally wounding their planning efforts. Thus, the continual preoccupation with process over methodology in this conference. In the eyes of many agency planners, the requirements for explicit risk/benefit analyses will complicate the process, introduce additional uncertainties into the decision making, increase costs, and reduce the potential for affirmative action on the final recommended

191

plan, as well as almost guaranteeing a significant extension of the
time required to bring a project from the point of initial definition of need to full operation. The reason for this is that risk/
benefit analyses will provide new opportunities and issues for use
by those anti-developmental groups dedicated to thwarting the proposed plans of the majority. The results of those analyses will
undoubtedly be used by these groups in their attempts to delay or
stop the project development process on the basis of additional
technicalities and/or additional environmental and public risk questions. It is my personal feeling (a feeling shared by others in
public water resources planning) that, if the process by which risk/
benefit analyses will be utilized is not improved, an immediate
impact of the new Principles, Standards and Procedures will be to
further extend the time gap between project conception and operation.

     The frustration and polarization generated by the current pro-
cedures, which effectively limit plan formulation to the responsible
agency and severely constrain meaningful dialogue between planners
and the affected public, must be minimized, and the energy now
expended on halting development plans must be channeled into improving them. There is a real need to develop explicit procedures to
insure real participation in the agency planning processes by all
affected parties on an equitable basis (as opposed to simply providing opportunities for observation and review).

     Separate discussion of the two aspects of risk/benefit analysis
--process and methodology--might prove more fruitful, since by no
means can the process questions be limited to risk and uncertainty.
One very important aspect which must be addressed under process is
the procedure for establishing acceptable levels of risk and uncertainty for the various segments of a water resources development
plan. Otherwise, the planners on the one hand will have no clear
guidelines and constraints, and on the other hand the reviewers
will be free to impose their own subjective criteria. It is not
difficult to foresee anti-developmentalists taking the position
(as they currently have in relation to nuclear energy development,
for example) that no level of risk is acceptable. The methodology
questions can certainly be discussed on the assumption, say, of an
equitable sharing of decision making authority, continued involvement of all decision makers throughout the planning process, and
agreements on the part of the decision makers to abide by the will
of the majority, while at the same time minimizing negative impacts
on the minorities. Such an approach should encourage a much more
positive and productive effort toward integration of the explicit
consideration and documentation of risk and uncertainty into all
aspects of water resources planning.

RATING THE RISKS*

Paul Slovic, Baruch Fischhoff, and Sarah Lichtenstein

Decision Research
A Branch of Perceptronics
Eugene, Oregon

People respond to the hazards they perceive. If their perceptions are faulty, efforts at public and environmental protection
are likely to be misdirected. In order to improve hazard management,
a risk assessment industry has developed over the last decade which
combines the efforts of physical, biological, and social scientists
in an attempt to identify hazards and measure the frequency and magnitude of their consequences.**

For some hazards extensive statistical data is readily available; for example, the frequency and severity of motor vehicle accidents are well documented. For other familiar activities, such as
the use of alcohol and tobacco, the hazardous effects are less readily discernible and their assessment requires complex epidemiological
and experimental studies. But in either case, the hard facts go only
so far and then human judgment is needed to interpret the findings
and determine their relevance for the future.

Other hazards, such as those associated with recombinant DNA
research or nuclear power, are so new that risk assessment must be
based on theoretical analyses such as fault trees (see Figure 1),
rather than on direct experience. While sophisticated, these analyses, too, include a large component of human judgment. Someone,

---

*This paper appeared in Societal Risk Assessment, How Safe is Safe
Enough?,R. Schwing & W. Albers, Jr., (Eds.) Plenum Press, NY, 1980.

**We have not attempted here to review all the important research in
this area. Interested readers should see Green [1978], Kates [1978],
and Otway, Maurer, and Thomas [1978].

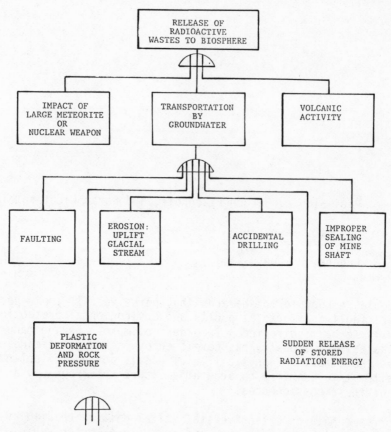

Fig. 1. Illustration of a fault tree. Fault trees are used most
often to characterize hazards for which direct experience
is not available. The tree shown here indicates the various ways in which radioactive material might accidentally
be released from nuclear wastes buried within a salt
deposit. To read this tree, start with the bottom row of
possible initiating events, each of which can lead to the
transportation of radioactivity by groundwater. This
transport can in turn release radioactivity to the bio-
sphere. As indicated by the second level of boxes,
release of radioactivity can also be produced directly
(without the help of groundwater) through the impact of
a large meteorite, a nuclear weapon, or a volcanic eruption. Fault trees may be used to map all relevant possibilities and to determine the probability of the final outcome. To accomplish this latter goal, the probabilities
of all component stages, as well as their logical connections, must be completely specified.
(Source: P. E. McGrath, 1974, Radioactive Waste Management,
Report EURFNR 1204, Karlsruhe, Germany)

relying on educated intuition, must determine the structure of the
problem, the consequences to be considered, and the importance of
the various branches of the fault tree.

Once the analyses have been performed, they must be communicated
to the various people who are actually responsible for dealing with
the hazards, including industrialists, environmentalists, regulators,
legislators, and voters. If these people do not see, understand, or
believe these risk statistics, then distrust, conflict, and effective
hazard management can result.

JUDGMENTAL BIASES

When lay people are asked to evaluate risks, they seldom have
statistical evidence on hand. In most cases they must rely on inferences based on what they remember hearing or observing about the risk
in question. Recent psychological research has identified a number
of general inferential rules that people seem to use in such situations [Tversky and Kahneman, 1974]. These judgmental rules, known
technically as heuristics, are employed to reduce difficult mental
tasks to simpler ones. Although valid in some circumstances, in
others they can lead to large and persistent biases with serious
implications for risk assessment.

Availability

One heuristic that has special relevance for risk perception
is known as "availability" [Tversky, Kahneman, 1973]. People who
use this heuristic judge an event as likely or frequent if instances
of it are easy to imagine or recall. Frequently occurring events
are generally easier to imagine and recall than rare events. Thus,
availability is often an appropriate cue. However, availability is
also affected by numerous factors unrelated to frequency of occurrence. For example, a recent disaster or a vivid film such as "Jaws"
can seriously distort risk judgments.

Availability-induced errors are illustrated by several recent
studies in which we asked college students and members of the League
of Women Voters to judge the frequency of various causes of death,
such as smallpox, tornadoes, and heart disease [Lichtenstein, Slovic,
Fischhoff, 1978]. In one study, these people were told the annual
death toll for motor vehicle accidents in the United States (50,000);
they were then asked to estimate the frequency of forty other causes
of death. In another study, participants were given two causes of
death and asked to judge which of the two is more frequent. Both
studies showed people's judgments to be moderately accurate in a
global sense; that is, people usually knew which were the most and

least frequent lethal events. However, within this global picture,
there was evidence that people made serious misjudgments, many of
which seemed to reflect availability bias.

Figure 2 compares the judged number of deaths per year with the
actual number according to public health statistics. If the frequency judgments were accurate, they would equal the actual death
rates, and all data points would fall on the straight line making
a 45 degree angle with the axes of the graph. In fact, the points
are scattered about a curved line that sometimes lies above and
sometimes below the line of accurate judgment. In general, rare
causes of death were overestimated and common causes of death were
underestimated. As a result, while the actual death toll varied
over a range of one million, average frequency judgments varied
over a range of only a thousand.

In addition to this general bias, many important specific
biases were evident. For example, accidents were judged to cause
as many deaths as diseases, whereas diseases actually take about
fifteen times as many lives. Homicides were incorrectly judged to
be more frequent than diabetes and stomach cancer. Homicides were
also judged to be about as frequent as stroke, although the latter
actually claims about eleven times as many lives. Frequencies of
death from botulism, tornadoes, and pregnancy (including childbirth
and abortion) were also greatly overestimated.

Table 1 lists the lethal events whose frequencies were most
poorly judged in our studies. In keeping with availability considerations, overestimated items were dramatic and sensational whereas
underestimated items tended to be unspectacular events which claim
one victim at a time and are common in nonfatal form.

In the public arena the availability heuristic may have many
effects. For example, the biasing effects of memorability and
imaginability may pose a barrier to open, objective discussions of
risk. Consider an engineer demonstrating the safety of subterranean
nuclear waste disposal by pointing out the improbability of each
branch of the fault tree in Figure 1. Rather than reassuring the
audience, the presentation might lead individuals to feel that "I
didn't realize there were so many things that could go wrong." The
very discussion of any low-probability hazard may increase the
judged probability of that hazard regardless of what the evidence
indicates.

In other situations, availability may lull people into compla-
cency. In a recent study [Fischhof, Slovic, Lichtenstein, 1978],
we presented people with various versions of a fault tree showing
the "risks" of starting a car. Participants were asked to judge
the completeness of the representation (reproduced in Figure 3).

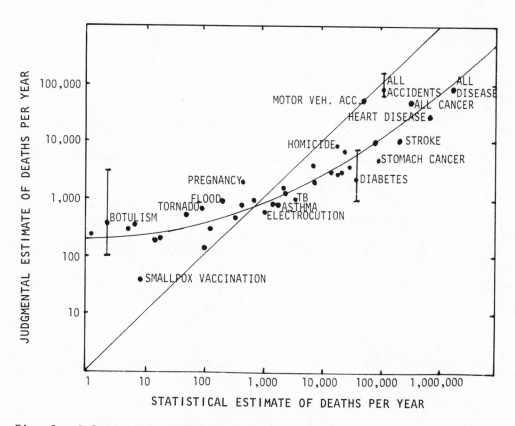

Fig. 2. Relationship between judged frequency and the actual number
         of deaths per year for 41 causes of death. If judged and
         actual frequencies were equal, the data would fall on the
         straight line. The points, and the curved line fitted to
         them, represent the averaged responses of a large number
         of lay people. While people were approximately accurate,
         their judgments were systematically distorted. As
         described in the text, both the compression of the scale
         and the scatter of the results indicate this. To give an
         idea of the degree of agreement among subjects, vertical
         bars are drawn to depict the 25th and 75th percentile of
         individual judgment for botulism, diabetes, and all acci-
         dents. Fifty percent of all judgments fall between these
         limits. The range of responses for the other 37 causes
         of death was similar. This figure is taken from Lichten-
         stein et al. [1978].

Table 1. Bias in Judged Frequency of Death

Most Overestimated	Most Underestimated
All accidents	Smallpox vaccination
Motor vehicle accidents	Diabetes
Pregnancy, childbirth, and abortion	Stomach cancer
Tornadœs	Lightning
Flood	Stroke
Botulism	Tuberculosis
All cancer	Asthma
Fire and flames	Emphysema
Venomous bite or sting	
Homicide	

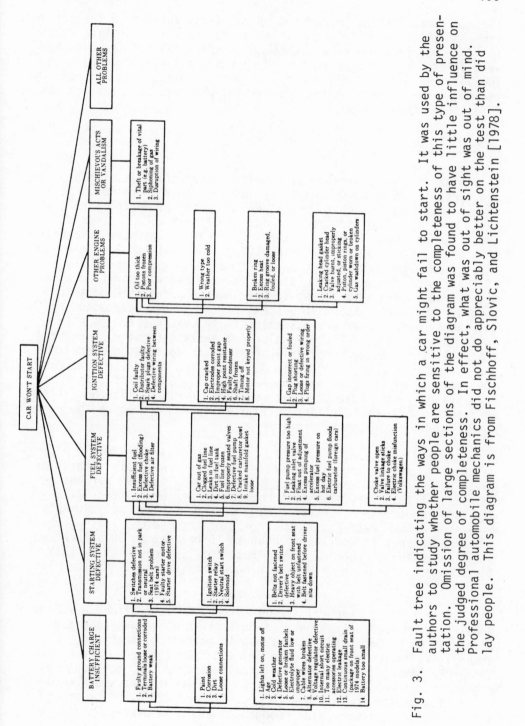

Fig. 3. Fault tree indicating the ways in which a car might fail to start. It was used by the authors to study whether people are sensitive to the completeness of this type of presentation. Omission of large sections of the diagram was found to have little influence on the judged degree of completeness. In effect, what was out of sight was out of mind. Professional automobile mechanics did not do appreciably better on the test than did lay people. This diagram is from Fischhoff, Slovic, and Lichtenstein [1978].

Their estimate of the proportion of no-starts falling in the category labeled "all other problems" was about the same when looking at the full tree of Figure 3 or at versions in which half of the branches were deleted. Such pruning should have dramatically increased the judged likelihood of "all other problems." However, it did not. In keeping with the availability heuristic, what was out of sight was effectively out of mind.

Overconfidence

A particularly pernicious aspect of heuristics is that people are typically very confident about judgments based on them. For example, in a follow-up to the study on causes of death, participants were asked to indicate the odds that they were correct in their judgment about which of two lethal events was more frequent [Fischhoff, Slovic, Lichtenstein, 1977]. Odds of 100:1 or greater were given often (25 percent of the time). However, about one out of every eight answers associated with such extreme confidence was wrong (fewer than 1 in 100 would have been wrong if the odds had been appropriate). About 30 percent of the judges gave odds greater than 50:1 to the incorrect assertion that homicides are more frequent than suicides. The psychological basis for this unwarranted certainty seems to be people's insensitivity to the tenuousness of the assumptions upon which their judgments are based (in this case, the validity of the availability heuristic). Such overconfidence is dangerous. It indicates that we often do not realize how little we know and how much additional information we need about the various problems and risks we face.

Overconfidence manifests itself in other ways as well. A typical task in estimating failure rates or other uncertain quantities is to set upper and lower bounds so that there is a 98 percent chance that the true value lies between them. Experiments with diverse groups of people making many different kinds of judgments have shown that, rather than 2 percent of true values falling outside the 98 percent confidence bounds, 20 percent to 50 percent do so [Lichtenstein, Fischhoff, Phillips, 1977]. People think that they can estimate such values with much greater precision than is actually the case.

Unfortunately, experts seem as prone to overconfidence as lay people. When the fault tree study described above was repeated with a group of professional automobile mechanics, they, too, were insensitive to how much had been deleted from the tree. Hynes and Vanmarcke [1976] asked seven "internationally known" geotechnical engineers to predict the height of an embankment that would cause a clay foundation to fail and to specify confidence bounds around this estimate that were wide enough to have a 50 percent chance of

enclosing the true failure height. None of the bounds specified by
these experts actually did enclose the true failure height. The
multi-million dollar Reactor Safety Study (the "Rasmussen Report")
[U.S. Nuclear Regulatory Commission, 1978], in assessing the probability of a core melt in a nuclear reactor, used a procedure for
setting confidence bounds that has been found in experiments to produce a high degree of overconfidence. Related problems led the
recent review committee, chaired by H. W. Lewis of the University
of California, Santa Barbara, to conclude that the Reactor Safety
Study greatly overestimated the precision with which the probability
of a core melt could be assessed [U.S. Nuclear Regulatory Commission,
1978].

    Another case in point is the 1976 collapse of the Teton Dam.
The committee on Government Operations has attributed this disaster
to the unwarranted confidence of engineers who were absolutely certain they had solved the many serious problems that arose during
construction [U.S. Government, 1976]. Indeed, in routine practice,
failure probabilities are not even calculated for new dams even
though about 1 in 300 fails when the reservoir is first filled.
Further anecdotal evidence of overconfidence may be found in many
other technical risk assessments. Some common ways in which experts
may overlook or misjudge pathways to disaster include:

- Failure to consider the ways in which human errors can affect
  technological systems. Example: The disastrous fire at the
  Brown's Ferry Nuclear Plant was caused by a technician checking
  for an air leak with a candle, in violation of standard operating procedures.

- Overconfidence in current scientific knowledge. Example: The
  failure to recognize the harmful effects of X-rays until societal use had become widespread and largely uncontrolled.

- Insensitivity to how a technological system functions as a
  whole. Example: Though the respiratory risk of fossil-fueled
  power plants has been recognized for some time, the related
  effects of acid rains on ecosystems were largely missed until
  very recently.

- Failure to anticipate human response to safety measures.
  Example: The partial protection offered by dams and levees
  gives people a false sense of security and promotes development
  of the flood plain. When a rare flood does exceed the capacity
  of the dam, the damage may be considerably greater than if the
  flood plain had been unprotected. Simiarly, "better" highways,
  while decreasing the death toll per vehicle mile, may increase
  the total number of deaths because they increase the number of
  miles driven.

Desire for Certainty

Every technology is a gamble of sorts and, like other gambles, its attractiveness depends on the probability and size of its possible gains and losses. Both scientific experiments and casual observation show that people have difficulty thinking about and resolving the risk/benefit conflicts even in simple gambles. One way to reduce the anxiety generated by confronting uncertainty is to deny that uncertainty. The denial resulting from this anxiety-reducing search for certainty thus represents an additional source of overconfidence. This type of denial is illustrated by the case of people faced with natural hazards, who often view their world as either perfectly safe or as predictable enough to preclude worry. Thus, some flood victims interviewed by Kates [1962] flatly denied that floods could ever recur in their areas. Some thought (incorrectly) that new dams and reservoirs in the area would contain all potential floods, while others attributed previous floods to freak combinations of circumstances, unlikely to recur. Denial, of course, has its limits. Many people feel that they cannot ignore the risks of nuclear power. For these people, the search for certainty is best satisfied by outlawing the risk.

Scientists and policy makers who point out the gambles involved in societal decisions are often resented for the anxiety they provoke. Borch [1968] noted how annoyed corporate managers get with consultants who give them the probabilities of possible events instead of telling them exactly what will happen. Just before a blue-ribbon panel of scientists reported that they were 95 percent certain that cyclamates do not cause cancer, Food and Drug Administration Commissioner Alexander Schmidt said, "I'm looking for a clear bill of health, not a wishy-washy, iffy answer on cyclamates" [Eugene Register-Guard, Jan. 14, 1976]. Senator Edmund Muskie has called for "one-armed" scientists who do not respond "on the one hand, the evidence is so, but on the other hand . . ." when asked about the health effects of pollutants [David, 1975].

The search for certainty is legitimate if it is done consciously, if the remaining uncertainties are acknowledged rather than ignored, and if people realize the costs. If a very high level of certainty is sought, those costs are likely to be high. Eliminating the uncertainty may mean eliminating the technology and foregoing its benefits. Often some risk is inevitable. Efforts to eliminate it may only alter its form. We must choose, for example, between the vicissitudes of nature on an unprotected flood plain and the less probable, but potentially more catastrophic, hazards associated with dams and levees.

ANALYZING JUDGMENTS OF RISK

In order to be of assistance in the hazard management process,
a theory of perceived risk must explain people's extreme aversion to
some hazards, their indifference to others, and the discrepancies
between these reactions and experts' recommendations. Why, for
example, do some communities react vigorously against locating a
liquid natural gas terminal in their vicinity despite the assurances
of experts that it is safe? Why do other communities situated on
flood plains and earthquake faults or below great dams show little
concern for the experts' warnings? Such behavior is doubtless
related to how people assess the quantitative characteristics of
the hazards they face. The preceding discussion of judgmental processes was designed to illuminate this aspect of perceived risk.
The studies reported below broaden the discussion to include more
qualitative components of perceived risk. They ask, when people
judge the risk inherent in a technology, are they referring only to
the (possibly misjudged) number of people it could kill or also to
other, more qualitative features of the risk it entails?

Quantifying Perceived Risk

In our first studies, we asked four different groups of people
to rate thirty different activities and technologies according to
the present risk of death from each [Fischhoff, Slovic, Lichtenstein,
Read, Combs, 1978; Slovic, Fischhoff, Lictenstein, 1978]. Three of
these groups were from Eugene, Oregon; they included 30 college students, 40 members of the League of Women Voters (LOWV), and 25 business and professional members of the "Active Club." The fourth
group was composed of 15 persons selected nationwide for their professional involvement in risk assessment. This "expert" group included a geographer, an environmental policy analyst, an economist,
a lawyer, a biologist, a biochemist, and a government regulator of
hazardous materials.

All these people were asked, for each of the thirty items,
"to consider the risk of dying (across all U.S. society as a whole)
as a consequence of this activity or technology." In order to make
the evaluation task easier, each activity appeared on a 3" × 5" card.
Respondents were told first to study the items individually, thinking of all the possible ways someone might die from each (e.g.,
fatalities from nonnuclear electricity were to include deaths
resulting from the mining of coal and other energy production activities as well as electrocution; motor-vehicle fatalities were to
include collisions with bicycles and pedestrians). Next, they were
to order the items from least to most risky and then assign numerical risk values by giving a rating of 10 to the least risky item
and making the other ratings accordingly. They were also given

additional suggestions, clarifications and encouragement to do as
accurate a job as possible. For example, they were told "A rating
of 12 indicates that that item is 1.2 times as risky as the least
risky item (i.e., 20 percent more risky). A rating of 200 means
that the item is 20 times as risky as the least risky item, to which
you assigned a 10. . . ." They were urged to cross-check and adjust
their numbers until they believed they were right.

Table 2 shows how the various groups ranked the relative riski-
ness of these 30 activities and technologies. There were many simi-
larities between the three groups of lay persons. For example, each
group believed that motorcycles, other motor vehicles, and handguns
were highly risky, and that vaccinations, home appliances, power
mowers, and football were relatively safe. However, there were
strong differences as well. Active Club members viewed pesticides
and spray cans as relatively much safer than did the other groups.
Nuclear power was rated as highest in risk by the LOWV and student
groups, but only eighth by the Active Club. The students viewed
contraceptives and food preservatives as riskier and swimming and
mountain climbing as safer than did the other lay groups. Experts'
judgments of risk differed markedly from the judgments of lay persons. The experts viewed electric power, surgery, swimming, and
X-rays as more risky than the other groups, and they judged nuclear
power, police work, and mountain climbing to be much less risky.

What Determines Risk Perception?

What do people mean when they say that a particular technology
is quite risky? A series of additional studies was conducted to
answer this question.

Perceived risk compared to frequency of death. When people
judge risk, as in the previous study, are they simply estimating
frequency of death? To answer this question, we collected the best
available technical estimates of the annual number of deaths from
each of the thirty activities included in our study. For some
cases, such as commercial aviation and handguns, there is good
statistical evidence based on counts of known victims. For other
cases, such as the lethal potential of nuclear or fossil-fuel
power plants, available estimates are based on uncertain inferences
about incompletely understood processes. For still others, such as
food coloring, we could find no estimates of annual fatalities.

For the 25 cases for which we found technical estimates for
annual frequency of death, we compared these estimates with perceived risk. Results for experts and the LOWV sample are shown in
Figure 4 (the results for the other lay groups were quite similar
to those from the LOWV sample). The experts' mean judgments were

Table 2. Ordering of Perceived Risk for
30 Activities and Technologies[a]

	Group 1 LOWV	Group 2 College Students	Group 3 Active Club Members	Group 4 Experts
Nuclear power	1	1	8	20
Motor vehicles	2	5	3	1
Handguns	3	2	1	4
Smoking	4	3	4	2
Motorcycles	5	6	2	6
Alcoholic beverages	6	7	5	3
General (private) aviation	7	15	11	12
Police work	8	8	7	17
Pesticides	9	4	15	8
Surgery	10	11	9	5
Fire fighting	11	10	6	18
Large construction	12	14	13	13
Hunting	13	18	10	23
Spray cans	14	13	23	26
Mountain climbing	15	22	12	29
Bicycles	16	24	14	15
Commercial aviation	17	16	18	16
Electric power	18	19	19	9
Swimming	19	30	17	10
Contraceptives	20	9	22	11
Skiing	21	25	16	30
X rays	22	17	24	7
High school & college football	23	26	21	27
Railroads	24	23	20	19
Food preservatives	25	12	28	14
Food coloring	26	20	30	21
Power mowers	27	28	25	28
Prescription antibiotics	28	21	26	24
Home appliances	29	27	27	22
Vaccinations	30	29	29	25

[a] The ordering is based on the geometric mean risk ratings within each group.
Rank 1 represents the most risky activity or technology.

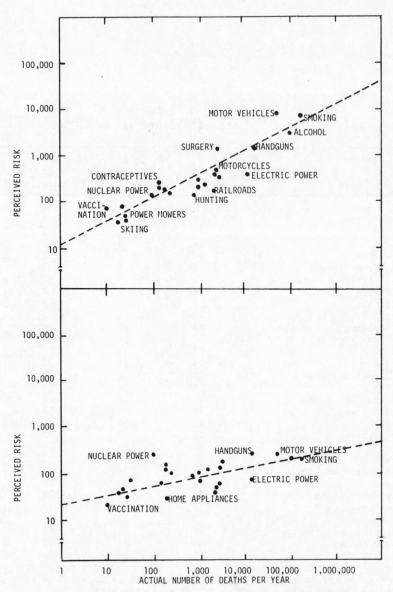

Fig. 4.   Judgments of perceived risk for experts (top) and lay people
          (bottom) plotted against the best technical estimates of
          annual fatalities for 25 technologies and activities. Each
          point represents the average responses of the participants.
          The dashed lines are the straight lines that best fit the
          points. The experts' risk judgments are seen to be more
          closely associated with annual fatality rates than are the
          lay judgments.

so closely related to the statistical or calculated frequencies that
it seems reasonable to conclude that they viewed the risk of an activity or technology as synonymous with its annual fatalities. The risk
judgments of lay people, however, showed only a moderate relationship
to the annual frequencies of death,* raising the possibility that,
for them, risk may not be synonymous with fatalities. In particular,
the perceived risk from nuclear power was disproportionately high
compared to its estimated number of fatalities.

Lay fatality estimates. Perhaps lay people based their risk
judgments on annual fatalities, but estimated their numbers inaccu-
rately. To test this hypothesis, we asked additional groups of students and LOWV members "to estimate how many people are likely to
die in the U.S. in the next year (if the next year is an average
year) as a consequence of these thirty activities and technologies."
We asked our student and LOWV samples to consider all sources of
death associated with these activities.

The mean fatality estimates of LOWV members and students are
shown in columns 2 and 3 of Table 3. If lay people really equate
risk with annual fatalities, one would expect that their own estimates of annual fatalities, no matter how inaccurate, would be very
similar to their judgments of risk. But this was not so. There was
a moderate agreement between their annual fatality estimates and
their risk judgments, but there were important exceptions. Most
notably, nuclear power had the lowest fatality estimate and the
highest perceived risk for both LOWV members and students. Overall,
lay people's risk perceptions were no more closely related to their
own fatality estimates than they were to the technical estimates
(Figure 4).

These results lead us to reject the idea that lay people wanted
to equate risk with annual fatality estimates but were inaccurate in
doing so. Instead, we are led to believe that lay people incorporate
other considerations besides annual fatalities into their concept of
risk.

Some other aspects of lay people's fatality estimates are of
interest. One is that they were moderately accurate. The relationship between the LOWV members' fatality estimates and the best technical estimates is plotted in Figure 5. The lay estimates showed
the same overestimation of those items that cause few fatalities and
underestimation of those resulting in the most fatalities that was

------

*The correlations between perceived risk and the annual frequencies
of death were .92 for the experts and .62, .50, and .56 for the
League of Women Voters, students, and Active Club samples, respectively.

Table 3. Fatality Estimates and Disaster Multipliers
          for 30 Activities and Technologies

Activity or Technology	Technical Fatality Estimates	Geometric Mean Fatality Estimates Average Year		Geometric Mean Multiplier Disastrous Year	
		LOWV	Students	LOWV	Students
1. Smoking	150,000	6,900	2,400	1.9	2.0
2. Alcoholic beverages	100,000	12,000	2,600	1.9	1.4
3. Motor vehicles	50,000	28,000	10,500	1.6	1.8
4. Handguns	17,000	3,000	1,900	2.6	2.0
5. Electric power	14,000	660	500	1.9	2.4
6. Motorcycles	3,000	1,600	1,600	1.8	1.6
7. Swimming	3,000	930	370	1.6	1.7
8. Surgery	2,800	2,500	900	1.5	1.6
9. X rays	2,300	90	40	2.7	1.6
10. Railroads	1,950	190	210	3.2	1.6
11. General (private) aviation	1,300	550	650	2.8	2.0
12. Large construction	1,000	400	370	2.1	1.4
13. Bicycles	1,000	910	420	1.8	1.4
14. Hunting	800	380	410	1.8	1.7
15. Home appliances	200	200	240	1.6	1.3
16. Fire fighting	195	220	390	2.3	2.2
17. Police work	160	460	390	2.1	1.9
18. Contraceptives	150	180	120	2.1	1.4
19. Commercial aviation	130	280	650	3.0	1.8
20. Nuclear power	100[a]	20	27	107.1	87.6
21. Mountain climbing	30	50	70	1.9	1.4
22. Power mowers	24	40	33	1.6	1.3
23. High school & college football	23	39	40	1.9	1.4
24. Skiing	18	55	72	1.9	1.6
25. Vaccinations	10	65	52	2.1	1.6
26. Food coloring	--[b]	38	33	3.5	1.4
27. Food preservatives	--[b]	61	63	3.9	1.7
28. Pesticides	--[b]	140	84	9.3	2.4
29. Prescription antibiotics	--[b]	160	290	2.3	1.6
30. Spray cans	--[b]	56	38	3.7	2.4

[a] Technical estimates for nuclear power were found to range between 16 and 600 annual fatalities. The geometric mean of these estimates was used here.

[b] Estimates were unavailable.

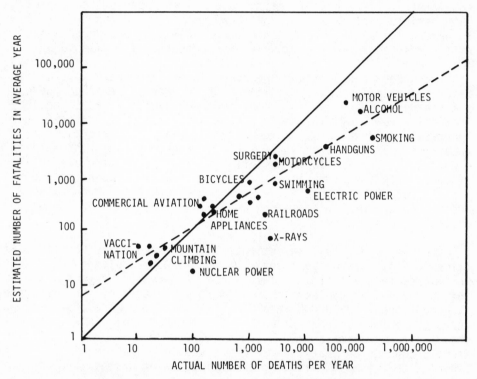

Fig. 5.   Lay people's judgments of the number of fatalities in an
average year plotted against the best estimates of annual
fatalities for 25 activities and technologies. The solid
line indicates accurate judgment, while the dashed line
best fits the data points. These results have much the
same character as those shown in Figure 2 for a different
collection of hazards. Low frequencies were overestimated
and high ones were underestimated. The overall relationship is marred by specific biases (e.g., the underestimation of fatalities associated with railroads, X-rays,
electric power, and smoking).

apparent in Figure 2 for a different collection of hazards. Also
as in Figure 2, the moderate overall relationship between lay and
technical estimates was marred by specific biases (e.g., the
underestimation of fatalities associated with railroads, X-rays,
electric power, and smoking).

Disaster potential. The fact that the LOWV members and students assigned very high risk values to nuclear power along with very low estimates of its annual fatality rates is an apparent contradiction. One possible explanation is that LOWV members expected nuclear power to have a low death rate in an average year but considered it to be a high risk technology because of its potential for disaster.

In order to understand the role played by expectations of disaster in determining lay people's risk judgments, we asked these same respondents to give a number for each activity and technology indicating how many times more deaths would occur if next year were "particularly disastrous" rather than average. The averages of these multipliers are shown in Table 3. For most activities, people saw little potential for disaster. For the LOWV sample all but five of the multipliers were less than 3, and for the student sample all but six were less than 2. The striking exception in both cases is nuclear power, with a geometric mean disaster multiplier in the neighborhood of 100.

For any individual an estimate of the expected number of fatalities in a disastrous year could be obtained by applying the disaster multiplier to the estimated fatalities for an average year. When this was done for nuclear power, almost 40 percent of the respondents expected more than 10,000 fatalities if next year were a disastrous year. More than 25 percent expected 100,000 or more fatalities. These extreme estimates can be contrasted with the Reactor Safety Study's conclusion that the maximum credible nuclear accident, coincident with the most unfavorable combination of weather and population density, would cause only 3,300 prompt fatalities [U.S. Nuclear Regulatory Commission, 1975]. Furthermore, that study estimated the odds against an accident of this magnitude occurring during the next year (assuming 100 operating reactors) to be about 2,000,000:1.

Apparently, disaster potential explains much or all of the discrepancy between the perceived risk and frequency of death values for nuclear power. Yet, because disaster plays only a small role in most of the thirty activities and technologies we have studied, it provides only a partial explanation of the perceived risk data.

Qualitative characteristics. Are there other determinants of risk perceptions besides frequency estimates? We asked experts, students, LOWV members, and Active Club members to rate the thirty technologies and activities on nine qualitative characteristics that have been hypothesized to be important [Lowrance, 1976]. These ratings scales are described in Table 4.

Examination of "risk profiles" based on mean ratings for the nine characteristics proved helpful in understanding the risk judgments of lay people. Nuclear power, for example, had the dubious

Table 4.   Risk Characteristics Rated by LOWV Members and Students

Voluntariness of risk

Do people face this risk voluntarily? If some of the risks are voluntarily
undertaken and some are not, mark an appropriate spot towards the center of
the scale.

risk assumed   1   2   3   4   5   6   7   risk assumed
voluntarily   involuntarily

Immediacy of effect

To what extent is the risk of death immediate--or is death likely to occur
at some later time?

effect   1   2   3   4   5   6   7   effect
immediate   delayed

Knowledge about risk

To what extent are the risks known precisely by the persons who are exposed
to those risks?

risk level   risk level
known   1   2   3   4   5   6   7   not
precisely   known

To what extent are the risks known to science?

risk level   risk level
known   1   2   3   4   5   6   7   not
precisely   known

Control over risk

If you are exposed to the risk, to what extent can you, by personal skill
or diligence, avoid death?

personal risk   personal risk
can't be   1   2   3   4   5   6   7   can be
controlled   controlled

Newness

Is this risk new and novel or old and familiar?

new 1   2   3   4   5   6   7   old

Chronic-catastrophic

Is this a risk that kills people one at a time (chronic risk) or a risk that
kills large numbers of people at once (catastrophic risk)?

chronic 1   2   3   4   5   6   7   catastrophic

Common-dread

Is this a risk that people have learned to live with and can think about
reasonably calmly, or is it one that people have great dread for--on the
level of a gut reaction?

common 1   2   3   4   5   6   7   dread

Severity of consequences

When the risk from the activity is realized in the form of a mishap or illness,
how likely is it that the consequence will be fatal?

certain   certain
not to be 1   2   3   4   5   6   7   to be
fatal   fatal

distinction of scoring at or near the extreme on all of the characteristics associated with high risk. Its risks were seen as involuntary, delayed, unknown, uncontrollable, unfamiliar, potentially
catastrophic, dreaded, and severe (certainly fatal). Its spectacular
and unique risk profile is contrasted in Figure 6 with non-nuclear
electric power and with another radiation technology, X-rays, both
of whose risks were judged to be much lower. Both electric power
and X-rays were judged much more voluntary, less catastrophic, less
dreaded, and more familiar than nuclear power.

Across all thirty items, ratings of dread and of the severity
of consequences were found to be closely related to lay person's
perceptions of risk. In fact, ratings of dread and severity along
with the subjective fatality estimates and the disaster multipliers
in Table 3 enabled the risk judgments of the LOWV and student groups
to be predicted almost perfectly.* Experts' judgments of risk were
not related to any of the nine qualitative risk characteristics.**

Judged seriousness of death. In a further attempt to improve
our understanding of perceived risk, we examined the hypothesis that
some hazards are feared more than others because the deaths they produce are much "worse" than deaths from other activities. We thought,
for example, that deaths from risks imposed involuntarily, from risks
not under one's control, or from hazards that are particularly
dreaded might be given greater weight in determining people's perceptions of risk.

However, when we asked students and LOWV members to judge the
relative "seriousness" of a death from each of the thirty activities
and technologies, the differences were slight. The most serious
forms of death (from nuclear power and handguns) were judged to be
only about two to four times worse than the least serious forms of
death (from alcoholic beverages and smoking). Furthermore, across
all thirty activities, judged seriousness of death was not closely
related to perceived risk of death.

_____

*The multiple correlation between the risk judgments of the LOWV mem-
bers and students and a linear combination of their fatality estimates, disaster multipliers, dread ratings, and severity ratings
was .95.

**A secondary finding was that both experts and lay persons believed
that the risks from most of the activities were better known to science than to the individuals at risk. The experts believed that the
discrepancy in knowledge was particularly great for vaccinations,
X-rays, antibiotics, alcohol, and home appliances. The only activities whose risks were judged better known to those exposed were
mountain climbing, fire fighting, hunting, skiing, and police work.

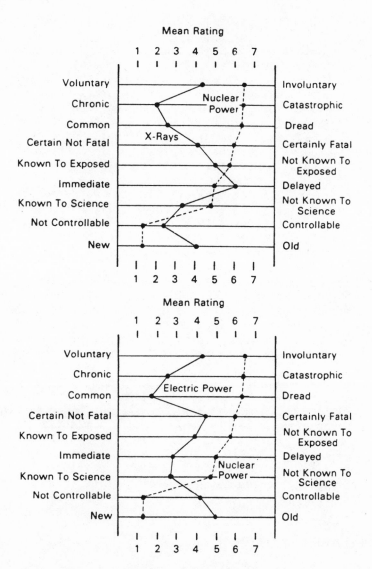

Fig. 6.   Qualitative characteristics of perceived risk for nuclear
          power and related technologies. In the top diagram, nuclear
          power and X-rays are compared. In the bottom diagram, risk
          profiles for nculear power and non-nuclear electric power
          are compared. Each profile consists of nine dimensions
          rated on a seven point scale. The instructions that elic-
          ited these responses are reproduced in Table 4. The per-
          ceived qualities of nuclear power are dramatically differ-
          ent from the comparison technologies. The source of this
          data is the LOWV sample studied by Fischhoff et al. [1978].

Reconciling Divergent Opinions

Our data show that experts and lay people have quite different perceptions about how risky certain technologies are. It would be comforting to believe that these divergent risk judgments would be responsive to new evidence so that, as information accumulates, perceptions would converge towards one "appropriate" view. Unfortunately, this is not likely to be the case. As noted earlier in our discussion of availability, risk perception is derived in part from fundamental modes of thought that lead people to rely on fallible indicators such as memorability and imaginability.

Furthermore, a great deal of research indicated that people's beliefs change slowly and are extraordinarily persistent in the face of contrary evidence [Ross, 1977]. Once formed, initial impressions tend to structure the way that subsequent evidence is interpreted. New evidence appears reliable and informative if it is consistent with one's initial belief; contrary evidence is dismissed as unreliable, erroneous, or unrepresentative. Thus, depending on one's predispositions, intense effort to reduce hazard may be interpreted to mean either that the risks are great or that the technologists are responsive to the public's concerns. Likewise, opponents of a technology may view minor mishaps as near catastrophes and dismiss the contrary opinions of experts as biased by vested interests.

From a statistical standpoint, convincing poeple that the catastrophe they fear is extremely unlikely is difficult under the best conditions. Any mishap could be seen as proof of high risk, whereas demonstrating safety would require a massive amount of evidence [Green and Bourne, 1972]. Nelkin's case history of a nuclear siting controversy [Nelkin, 1974] provides a good example of the inability of technical arguments to change opinions. In that debate each side capitalized on technical ambiguities in ways that reinforced its own position.

THE FALLIBILITY OF JUDGMENT

Our examination of risk perception leads us to the following conclusions:

- Cognitive limitations, coupled with the anxieties generated by facing life as a gamble, cause uncertainty to be denied, risks to be distorted, and statements of fact to be believed with unwarranted confidence.

- Perceived risk is influenced (and sometimes biased) by the imaginability and memorability of the hazard. People may, therefore, not have valid perceptions even for familiar risks.

- Our experts' risk perceptions correspond closely to statistical frequencies of death. Lay people's risk perceptions were based in part upon frequencies of death, but there were some striking discrepancies. It appears that for lay people, the concept of risk includes qualitative aspects such as dread and the likelihood of a mishap being fatal. Lay people's risk perceptions were also affected by catastrophic potential.

- Disagreements about risk should not be expected to evaporate in the presence of "evidence." Definitive evidence, particularly about rare hazards, is difficult to obtain. Weaker information is likely to be interpreted in a way that reinforces existing beliefs.

The significance of these results hinges upon one's acceptance of our assumption that subjective judgments are central to the hazard management process. Our conclusions mean little if one can assume that there are analytical tools which can be used to assess most risks in a mechanical fashion and that all decision makers have perfect information and the know-how to use it properly. These results gain in importance to the extent that one believes, as we do, that expertise involves a large component of judgment, that the facts are not all in (or obtainable) regarding many important hazards, that people are often poorly informed or misinformed, and that they respond not just to numbers but also to qualitative aspects of hazards.

Whatever role judgment plays, its products should be treated with caution. Research not only demonstrates that judgment is fallible, but it shows that the degree of fallibility is often surprisingly great and that faulty beliefs may be held with great confidence.

When it can be shown that even well-informed lay people have difficulty judging risks accurately, it is tempting to conclude that the public should be removed from the hazard-management process. The political ramifications of such a transfer of power to a technical elite are obvious. Indeed, it seems doubtful that such a massive disenfranchisement is feasible in any democratic society.

Furthermore, this transfer of decision making would seem to be misguided. For one thing, we have no assurance that experts' judgments are immune to biases once they are forced to go beyond their precise knowledge and rely upon their judgment. Although judgmental biases have most often been demonstrated with lay people, there is evidence that the cognitive functioning of experts is basically like that of everyone else.

In addition, in many if not most cases effective hazard management requires the cooperation of a large body of lay people. These people must agree to do without some things and accept substitutes for others; they must vote sensibly on ballot measures and for legislators who will serve them as surrogate hazard managers; they must obey safety rules and use the legal system responsibly. Even if the experts were much better judges of risk than lay people, giving experts an exclusive franchise on hazard management would involve substituting short-term efficiency for the long-term effort needed to create an informed citizenry.

For those of us who are not experts, these findings pose an important series of challenges: to be better informed, to rely less on unexamined or unsupported judgments, to be aware of the qualitative aspects that strongly condition risk judgments, and to be open to new evidence that may alter our current risk perceptions.

For the experts, our findings pose what may be a more difficult challenge: to recognize their own cognitive limitations, to temper their assessments of risk with the important qualitative aspects of risk that influence the responses of lay people, and somehow to create ways in which these considerations can find expression in hazard management without, in the process, creating more heat than light.

ACKNOWLEDGMENTS

The authors wish to express their appreciation to Christoph Hohenemser, Roger Kasperson, and Robert Kates for their many helpful comments and suggestions. This work was supported by the National Science Foundation under Grant ENV77-15332 to Perceptronics, Inc. Any opinions, findings and conclusions, or recommendations expressed herein are those of the authors and do not necessarily reflect the views of the National Science Foundation

REFERENCES

Borch, K., 1968, "The Economics of Uncertainty," Princeton University Press, Princeton, N.J.
David, E. E., 1975, One-armed scientists? Science, 189:891.
Eugene Register-Guard, Jan. 14, 1976, Doubts Linger on Cyclamate Risks.
Fischhoff, B., Slovic, P., and Lichtenstein, S., 1977, Knowing with certainty: The appropriateness of extreme confidence, J. Exp. Psych.: Human Perception and Performance, 3:552-564.

Fischhoff, B., Slovic, P., and Lichtenstein, S., 1978, Fault trees:
     Sensitivity of estimated failure probabilities to problem
     representation, J. Exp. Psych.: Human Perception and Perfor-
     mance, 4:342-355.
Fischhoff, B., Slovic, P., Lichtenstein, S., Read, S., and Combs, B.,
     1978, How safe is safe enough? A psychometric study of atti-
     tudes towards technological risks and benefits, Policy Sci.,
     8:127-152.
Green, A. E., and Bourne, A. J., 1972, "Reliability Technology,"
     Wiley Interscience, New York.
Green, C. H., 1978, Risk: Attitudes and beliefs, in: D. V. Canter,
     ed., "Behavior in Fires," forthcoming.
Hynes, M., and Vanmarcke, E., 1976, Reliability of embankment per-
     formance predictions, Proceedings, ASCE Engineering Mechanics
     Division Specialty Conference, University of Waterloo Press,
     Waterloo, Ontario, Canada.
Kates, R. W., 1962, Hazard and choice perception in flood plain
     management, Research Paper 78, Department of Geography, Univer-
     sity of Chicago, Chicago.
Kates, R. W., 1978, "Risk Assessment of Environmental Hazard," Wiley,
     New York.
Lichtenstein, S., Fischhoff, B., and Phillips, L. D., 1977, Calibra-
     tion of probabilities: The state of the art, in: H. Jungermann
     and G. de Zeeuw, eds., "Decision Making and Change in Human
     Affairs," Reidel, Dordrecht, The Netherlands.
Lichtenstein, S., Slovic, P., Fischhoff, B., Layman, M., and Combs,
     B., 1978, Judged frequency of lethal events, J. Exp. Psych.:
     Human Learning and Memory, 4:551-578.
Lowrance, W., 1976, "Of Acceptable Risk," Kaufman, Los Altos, Calif.
Nelkin, D., 1974, The role of experts on a nuclear siting contro-
     versy, Bull. Atomic Scientists, 30:29-36.
Otway, H. J., Maurer, D., and Thomas, K., 1978, Nuclear power:
     The questions of public acceptance, Futures, April.
Ross, L., 1977, The intuitive psychologist and his shortcomings,
     in: L. Berkowitz, ed., "Advances in Social Psychology,"
     Academic Press, New York.
Slovic, P., Fischhoff, B., and Lichtenstein, S., Expressed prefer-
     ences, unpublished manuscript, Decision Research, Eugene, Ore.
Tversky, A., and Kahneman, D., 1973, Availability: A heuristic for
     judging frequency and probability, Cognitive Psych., 4:207-232.
Tversky, A., and Kahneman, D., 1974, Judgment and uncertainty:
     Heuristics and biases, Science, 185:1124-1131.
U.S. Government, 1976, Teton dam disaster, Committee on Government
     Operations, Washington, D.C.
U.S. Nuclear Regulatory Commission, Oct. 1975, Reactor safety study:
     An assessment of accident risks in U.S. commercial nuclear
     power plants, WASH 1400, NUREG-75/014, Washington, D.C.
U.S. Nuclear Regulatory Commission, Sept. 1978, Risk assessment
     review group report to the U.S. Nuclear Regulatory Commission,
     NUREG/CR-0400.

RISK ASSESSMENT: ARID AND SEMIARID LANDS PERSPECTIVE

J. Eleonora Sabadell

Bureau of Land Management
Department of Interior
Washington, D.C.

The water crisis facing the Western U.S. may be as serious, if
not more so, than the energy crisis, because water is critical and
limiting to any further growth and development in our western and
Great Plains drylands.

In the U.S., ten percent of the developed cropland (or about
42 million acres) are irrigated, most of them located in arid and
semiarid areas. During 1975, 57 bgd (billion gallons per day) of
ground water were used for irrigation in the country. Exact figures
are lacking, but probably 40% of the total irrigated agriculture in
the western states depends on ground water. Taking into account
that agricultural exports are in the range of $20 to $25 billion per
year, with particular interest in specialty crops, it is evident that
any detrimental change in the supply of the water and the conditions
of the soil needed to maintain such activities could have serious
consequences.

In the U.S. today 7.5 bgd of water, or 2.5% of all withdrawals,
are used by minerals industries (metals, nonmetals, and fuels), and
water withdrawals for fuel mining and processing constitute about
36% of the total withdrawn by the mineral industry, but the consumptive use is 62% of the total mineral mining water consumption. The
coal and oil shale mining and processing are and will be mainly
located in arid-semiarid regions, a very important one being the
Upper Colorado region, creating heavy competition between this and
other interests for the allocation of water resources. In 1975,

The author wishes to acknowledge the substantive help of staff
member Becky Thornton.

ground water constituted over 40% of the total fresh water withdrawals (approximately 82 bgd) and water mining is increasing at
an accelerated rate, largely due to agricultural and settlement
uses. During that year overdrafts in the Texas region alone amounted
to 78 percent of the withdrawals, 50 percent in the Lower Colorado
region and 30 percent in the Rio Grande region. In central Arizona,
ground water levels are declining from 7 to 10 feet per year. The
future facing these regions is indeed uncertain, given the high probability of rapid ground water depletion, even if the rate of water
mining is maintained at the present levels.

Aside from the expected population growth in the U.S., an
internal migration from the snow belt to the sun belt is occurring
also at an increasing rate, with a popular preference for the dry,
warm climate of central and southwestern states. One impact of this
trend is that a projection for the year 2000 is of an increment in
water withdrawals and in the water-consumptive use of up to 50% of
the 1975 values for the Colorado, California, Rio Grande, Great
Basin and Texas Gulf regions. Today the total freshwater withdrawals in these regions are between 80 and 90 bgd, representing 26% of
the total withdrawals in the nation, 60% of which come from surface
water. The projected more-intensive use and reuse of the available
surface and ground water supplies present quality problems that
will become critical, requiring growing capital outlays and improved
technology to solve. It is estimated that by the year 2000, there
will be a deficiency in water supply for California's urban and agricultural demands totaling over two million acre-feet per year
[Gertsch, 1977]. These figures show in some measure the intense
interaction between the water resources of arid-semiarid regions,
other natural resources, human activity, and the far-reaching consequences of changes in the existing water supply-demand system.

It is evident that politicians and decision makers will be, as
some already are, facing difficult choices in the allocation of a
scarce resource to conflicting uses and interests. Risk analysis
may become an indispensable tool in selecting alternatives, but the
risk/benefit specialists must realize the special characteristics
of this type of decision.

A brief description of some main problems the drylands water
resources present in planning and management follows.

SURFACE WATER PROBLEMS

Surface water supplies are unevenly distributed over the western states and within each state itself. Further, these supplies
vary considerably from year to year and in different seasons of the
year. Because of these wide variations in distribution and runoff

patterns, in many areas of the West economic development has been dependent on construction of storage and conveyance facilities to provide adequate water supplies.

Water programs have primarily emphasized measures to offset this variability by seeking means of controlling and regulating the natural streamflows, both to reduce flooding during periods of high flow and to provide water for desired uses during periods of low flow or high withdrawals. Similarly, the differences in precipitation and runoff between areas have encouraged residents of relatively low-flow areas to seek water transfers from areas where flows were greater or where the water was thought to be less fully used. In response, governmental--federal, state, or local--and nongovernmental entities have made major investments in water control, storage, transfer, and distribution works. As shown in Table 1, a change in water supply can have a great impact on other economic variables.

Surface water quality also varies widely throughout the West. The upstream reaches of streams are generally of excellent quality. The downstream reaches of major streams, terminal lakes, estuaries, and coastal waters are subject to quality degradation caused by pollutants from municipal and industrial discharges, agricultural activities, resources development and utilization, and natural sources. The sources of water pollution may be grouped into two broad categories: point sources and nonpoint sources.

A chief characteristic of point sources is that pollutants from them are discharged into waters at identifiable and discrete points. The pollutants from nonpoint sources may originate from numerous sites scattered throughout a given area and are transported to surface and groundwaters through diffuse routes. These sources are especially significant in the West. The pollutants include sediments, natural salts, pesticides, chemical fertilizers, animal wastes, plant residues, salts, minerals, oil, acid, and numerous other substances. They find their way into water through diffuse overland runoff from rural and urban areas, seepage, natural drainage channels, man-made drainage systems, and various combinations of these routes.

Erosion and sedimentation both from natural and man-made sources are common problems throughout the West and are greatly accelerated where man's activity has modified the vegetative cover. Sediment, the product of erosion, may cause damage during transport all along streams, rivers, lakes, and wherever it is deposited. Sediment can result in overwash, swamping, and increased flooding. It accumulates in reservoirs, increases treatment costs of municipal and industrial supplies, makes navigable streams impassible without dredging, clogs irrigation and drainage improvements, smothers growing plants and spoils harvestable crops, increases maintenance costs of utility and transportation facilities, decreases the

Table 1. Effects of a Water Supply Increase or Decrease of One
Million Acre-Feet on Texas Statewide Economic Variables

Selected Sectors	Total Output*	Employment (Jobs)	Gain or Loss Personal Income*	Tax Collections*	Value Added in Production*
Agricultural crop sectors					
Irrigated cotton	543.67	14,861	156.85	51.01	310.54
Irrigated food grains	228.15	6,316	63.83	21.23	138.91
Irrigated feed grains	214.15	6,831	63.73	20.37	130.27
Other irrigated crops	330.13	11,491	89.80	34.54	209.96
Livestock sectors					
Range livestock	12.77	436	3.61	1.19	7.61
Feedlot livestock	213.10	4,895	44.79	15.41	97.46
Dairy farms	129.17	3,712	31.38	11.97	73.13
Poultry farms	180.10	4,580	36.90	15.65	83.83
Manufacturing sectors					
Processed food prod.	83.37	1,795	15.57	8.02	47.49
Agric. chemicals	15.74	287	2.87	1.55	10.80
Petroleum refining	58.28	997	10.80	5.61	36.95
Aircraft & parts	79.89	2,051	22.61	8.39	45.59
The municipal sector	6.18	244	1.78	0.53	3.91
Mining sectors					
Crude petroleum and natural gas	26.25	504	6.60	3.90	22.93
Other mining and quarrying	6.13	160	1.62	0.91	5.15
The electric svcs. sector	12.14	262	2.48	1.97	10.83

*$ millions; federal, state, local, and education.

Source: Texas Water Development Board, "Continuing Water Resources Planning and Development for Texas," Vol. 1, May 1977.

recreational value of water, and adversely affects the fishery
resource.

Over four billion tons of soil are picked up each year by the
nation's streams, and 25% of this total tonnage is discharged into
the oceans. The other 75% is moved around within drainage basins.
The dollar amount of damages caused by sediment is difficult to
determine, but it has been estimated at over $500 million per year.

The concentration of dissolved solids in the streamflow is
also of great concern. This concentration occurs partly because
some of the water withdrawn for such uses as irrigation returns to
the stream bearing dissolved material. Also, when water is evapor-
ated from reservoirs the salts remain and their concentration is
increased. Any diversion of flow leaves less water to dilute high
salt concentrations that result from natural causes, such as salt
springs. Consumptive losses associated with municipal and industrial water uses, evapotranspiration losses from native vegetation
on noncropped land, and out-of-basin diversions of water are other
causes of increased salinity even if no more salt is added to the
system. The salinity problem is increasing not only in surface
water but in ground waters as well. Salinity problems are in general related to poor drainage, inappropriate irrigation practices,
and disturbances of saline geological formations and soils. Salinity affects agricultural productivity, and about 25% of the irrigated land in the U.S. suffers from some degree of salinization and/
or waterlogging.

The Colorado River is an example of quality problems caused
by dissolved solids. Generally, flows in the headwaters of the
Colorado River are of high quality, usually with less than 50 parts
per million (ppm) of dissolved solids, but the concentration
increases progressively downstream from both natural and man-made
causes to about 20-fold. At Imperial Dam in the Lower Colorado
River Basin, the measured flows have had an average concentration
of about 865 ppm. If present trends continue, it is estimated that
the concentration will reach 1340 ppm by the year 2000, two-and-a-
half times the recommended maximum allowable concentration for
municipal water supplies, and dangerously high for agricultural use
[NWC, 1973]. This has reduced Salton Sea fisheries production, the
productivity of the land base, and the palatability of water supplies for domestic use. It has added treatment costs for municipal
supplies, increased corrosion and scaling of pipes and equipment,
accelerated fabric wear, and added treatment costs of industrial
water and other costs for industrial makeup and processed water.
It has also caused significant reductions in U.S. and Mexican agricultural production. It has been estimated that the damage to agriculture is about $40,000 annually per mg/l salt load and that muni-
cipal damages are about $289,000 per mg/l. In the San Joaquin

Table 2.   Salinity Damages in the Lower Colorado River Basin

1975 Dollars

	Estimated Average $/mg/l	1977 Estimated Damage @ 6.7% Inflation Rate $/mg/l
Agricultural damages	$ 33,168	$ 40,000
Municipal (household) damages		
Metropolitan water district	187,000	
Central Arizona service area	26,300	
Lower main stem	27,200	
	$240,500	$289,000
Industrial damages	1,500	1,800
Average agricultural, municipal and industrial damages estimated from salinity		$330,800

Source: Bureau of Land Management, "The Effects of Surface Distur-
bance on the Salinity of Public Lands in the Upper Colorado
River Basin" (1977 Status Report), February 1978.

Valley more than $30 million have been lost yearly as a result of
high saline water tables. If irrigation proceeds there in the same
fashion today, in twenty years around 3/4 of a million acres will be
affected, with an annual loss of $300 million [Grew, 1979]. If we
consider that today there are more than 34 million acres of irrigated cropland in the 17 western states, the magnitude of the potential problem, if irrigation is continued without proper safeguards,
can be surmised.

Fertilizer and pesticide use has expanded rapidly in the 17
western states during the last thirty years, in order to maximize
crop production. This in turn has impacted heavily on the surface
and ground water resources. Under Section 208 of the Federal Water
Pollution Control Act amendments of 1972--which concern runoff,
agricultural, urban and any other nonpoint pollution sources--pro-
cedures and methods of control will be adopted. States have completed some management plans, but the two-year time frame has been

recognized as insufficient. Amendments were enacted in 1977 to
remedy this, and Section 208 planning has been completed.

GROUND WATER PROBLEMS

     Mining of ground water occurs when rates of withdrawal exceed
the net recharge; and when mining continues over time, on a sustained
basis, several things may happen: ground water tables decline,
making the pumping of water increasingly expensive; compaction may
occur in the aquifer, adversely affecting storage capacity and trans-
missivity; quality may be threatened by salt water intrusion; and
subsidence may occur.

     A prime example of ground water mining in an aquifer system with
negligible recharge is found in the Ogallala Formation in the High
Plains of Texas, where the surface water resources are also limited.
Most of the irrigated acreage has been developed since World War II.
The population of the area has increased in the major towns and
cities and the area is heavily dependent on irrigation. Pumpage
from 1953 to 1961 averaged 5 million acre-feet annually, compared
to estimates of annual recharge of only 100,000 to 350,000 acre-
feet. In Nebraska alone between 4 and 5 million acres today are
irrigated with approximately 50,000 center-pivot systems. This
number keeps increasing annually. As a result, the resource is
being rapidly depleted. By 2015, irrigated acreage, without other
sources of water, is projected to decline to 125,000 acres and water
pumpage to 95,000 acre-feet annually. A similar situation is developing in other portions of the Ogallala Formation, which extends as
far north as the Platte River and underlies portions of New Mexico,
Oklahoma, eastern Colorado, western Kansas, and Nebraska.

     Ground water quality is also changing and concentrations of
nitrates, fluoride, and total dissolved solids are increasing to the
point where damaging levels could be reached. Because they are more
lasting, the effects of ground water pollution can be more significant than the effects of surface water pollution, yet, paradoxically,
less attention has been paid to ground water pollution and less is
known about it. There is a great variety in the sources of ground
water pollution. Some are obvious, such as waste disposal wells
into which toxic substances are intentionally injected. Others are
unintended by-products of other activities: oil and gas operations
which permit the seepage of petroleum or salt water into freshwater
strata; agricultural operations which permit introduction of excess
applications of fertilizers and pesticides into the ground water;
percolation of saline water into fresh water aquifers. In some
instances, the occurrence of contamination is not known immediately
because time often elapses before the effects become evident.

WATER AND ENERGY

Water and energy are closely tied since energy is needed to develop and maintain water supply and distribution systems, for water quality control processes, and for wastewater treatment systems. On the other hand, water is needed to extract and process fuels, to produce energy, and for cooling. Water and energy are necessary together for food and fiber production, industrialization, urbanization; in short, for fulfilling all human and environmental needs [Fairchild, 1973].

The role of western water in meeting high-priority energy needs is tied primarily to the development of large reserves of coal and oil shale, to the disposal of waste heat from thermal electric and fuel conversion plants, and to supplying associated urban growth.

Water will be required for any process that utilizes these resources, whether it be mine-mouth coal electric plants, plants to convert coal to gas or liquid fuels, or even mining and shipping the coal to markets outside the basin. In addition to water needed directly in the mining and energy conversion process, water will be needed for residential use by the expanded local population and also may be needed, at least initially, to reclaim strip-mined lands. If we develop these energy sources to any large extent, the water supplies must be carefully evaluated, developed, and conserved to assure that our energy needs, as well as the other water needs such as irrigation, municipal and industrial supplies, fish and wildlife, and recreation, are also met. If the energy sources of the Missouri and Colorado River areas are to be developed in time to prevent severe economic dislocations because of energy shortages, an early start must be made to plan and develop the water supplies. Timing is critical because a much larger lead-time is usually needed to develop and convey water than is normally required to develop energy.

However, some feel that the level of synfuels development does not appear to be constrained by the volume of water available. Throughout the western states, the general view is that impediments to development projects will be due to factors other than water availability--such as air quality, undesirable boom towns or other social considerations, outdated technology, and project economics.

Another rough consensus is that western agriculture is not imperiled by the demands of reasonably foreseeable energy projects. On a state-by-state basis this may be true, although in some local specific situations marginal agriculture may be displaced and in others greater efficiency in use, including increased local storage, will be necessary.

These general views rest upon the belief that energy development in the western states will come nowhere near approximating the levels of the earlier projections mentioned above, particularly in the development of synthetic fuels from coal and oil shale. Geothermal and other potential energy sources are not perceived as causing great demands on the region's water resources before the turn of the century. Increased export of coal to areas outside the region for conversion to electric power may lessen the need for mine-mouth power plants, but a strong national commitment to a subsidized syn-fuels development program would unquestionably raise concerns, not unanimous throughout the West, over the ability of some river basins to sustain development consistent with present and future needs of agriculture, recreation, and municipalities.

Another consideration in the water-energy relationship is that energy resource potentials involve extensive areas of publicly owned lands and a variety of environmental, social, and economic impacts are of vital concern to local, state, and national interests. Coordinated regional planning and organizational arrangements for advance selection of energy sites, more efficient approval procedures at all levels of government, and improved long-range planning, are important steps in integrating good water and land resource management with the conservation and use of the energy assets of the West.

WATER ON INDIAN RESERVATIONS

There is a great deficiency in studies of resource potentials and the development of water programs and projects for a large number of Indian reservations across the 17 western states. Identification and quantification of water requirements are requisite to the self-determination of the Indians' future.

Indians have both economic and cultural ties to the land, and much of their economy is based on livestock grazing. This type of land use, sometimes coupled with the aridity of the area and inappropriate resources management, has resulted in localized overgrazing, leading to locally severe water and soil problems.

Another area of concern is related to the available procedures to accommodate both Indian and non-Indian water needs. In water-short areas, some of the competing demands and Indian entitlements to water will have to be settled by court decisions and orders. Conflict between Indian and non-Indian uses of water arise from the fact that many non-Indian water resource projects rely on supplies in which Indians have water rights with earlier priorities. For example, the Navajo claim that Winters Doctrine rights have a higher priority on the Little Colorado River than present users and that present users may have to yield to Navajo rights in the near future.

The new BIA water policy seeks to use negotiation where possible instead of litigation to settle Indian water claims and to develop and initiate beneficial use of water at the earliest opportunity. The policy seeks to accomplish these goals by (1) changing to criteria more appropriate to the extenuating circumstances of the American Indians on which to base decisions of water use, and (2) by completing an assessment of current availability and future needs for water on which to base negotiations. Negotiation appears to be more advantageous because litigation is time-consuming and these delays may enable others to develop water that may be Indian water, and also because litigation is more expensive and the money may be better applied in the development of water resources.

WATER RIGHTS

Water rights serve as absolute constraints on the way in which both water and land are used, now as in the past. Once established in the law by statute, common law, or customary practice, these systems tend to resist change and are especially difficult to replace or modernize. As competition for limited water supplies increases, the water rights systems inherited from the past sometimes hamper desirable water resource management reforms [Templer, 1978a].

In the humid eastern United States, property owners along a stream or river enjoy riparian rights, which gives equal water use rights to all owners, whether use is made or not. In the arid western states, riparian rights made little sense because water rarely appeared in the channels. It was more important that the water be used when available, even if the user was not located along the watercourse. Thus, one who appropriated water for his own use obtained the right to its continued use, in the same general amounts and for the same purposes. One who failed to continue such use lost his appropriative rights by abandonment. On the other hand, as long as such use continued, the rights created thereby were superior to those of any subsequent user. Moreover, the water rights could be transferred separately from the land, but the transfer of land generally carried no special rights to waters in abutting channels. In addition, water could be appropriated by communities as well as by individuals, and this practice was expanded to include all levels of government, including the federal government. Appropriations today are controlled by comprehensive state and federal laws and regulations which prescribe higher priorities for certain uses than others, and also control the use and disposition of the waters.

The major provisions of the appropriation doctrine developed out of necessity to allow for orderly growth and development in the arid West. They are:

1. It is the beneficial use, not the place or ownership of the land, that determines the right to use of the water. Diversions for such varied uses as mining, irrigation, municipal, and industrial supplies are possible.

2. The amount of water appropriated does not vary with the amount in the stream. A definite quantity of water is appropriated.

3. In times of water shortage, it is the time priority of use, not equality of right, that determines use. There is generally no proration.

4. Appropriation of water is a property right. As long as the water is used beneficially, the water right continues. However, appropriation water rights may be sold or transferred [Trelease, 1976].

In many parts of the United States the only available water was underground water. The courts in some of these states, therefore, extended principles comparable to appropriative water rights to ground water. Whoever first used water from beneath the ground, even though not on his property, obtained the right to continue such use [Howard, 1978]. Some other states have adopted different methods of regulating their ground waters; still others have not adopted any comprehensive laws regarding regulation of ground water.

Some regulations concerning well spacing, water proration, conservation, and off-farm ground-water waste are now going into effect in some states that seek to avoid problems of water mining. Considering the substantial interconnections which can exist between water in the various phases of the hydrologic cycle, it is generally agreed that conjunctive use and management of water resources is desirable, especially in situations where it can be demonstrated that uncontrolled water use in one phase of the cycle can have an appreciable effect on established surface water rights through reduction in base flow of streams [Templer, 1978b].

Because such a large portion of the western states is public land, the issue of water rights on public land must be addressed as well. The National Park Service, the Fish and Wildlife Service, the Water and Power Resources Service (formerly Bureau of Reclamation), the Forest Service, and the Bureau of Land Management are the major agencies of the federal government which have significant amounts of land which require water. The federal government claims both reserved water rights and non-reserved appropriated water rights on its reserved and public lands. However, the latter right will have to be tested in the courts and has been criticized by the western states. The development of methods to inventory and quantify the federal government's water rights and integrate them into the various state systems is part of the June 1979 draft report of the

Federal Task Force on Federal (Non-Indian) Reserved Water Rights, which report is being prepared. Federal water rights for Indian lands represent a much larger claim, quantitatively, than non-Indian rights. Indian rights are a trust responsibility of the federal government and can be settled only in close consultation with individual Indian tribes.

In general, the legal availability of water, in contrast to its physical availability, presents more problematic constraints. In practice, legal and institutional controls peculiar to each state restrict the amount of water to which rights may be purchased, and massive transfers of water rights to newer or expanded uses may need to be determined more and more in the political arena in each state.

CONCLUSIONS

Some of the water-related problems have been briefly described to illustrate their variety and complexity, their interrelatedness, and the emergence of new issues in allocating, planning, and managing our drylands water resources that must be addressed. It should be recognized that changes are occurring in the decision-making process, where new and expanding interests are applying their leverage to the political process, and where newcomers, such as the executive and judiciary powers, are increasingly participating in regulatory and legislative matters. It is also becoming apparent that more emphasis must be given to the long-term consequences of decisions made and policies adopted. Further, the attention given eminently to the supply of water should shift, at least in part, to the demand part of the equation, seeking means to better accommodate changes and expansions in arid and semi-arid land uses.

It is evident that the resilience in some of our drylands is or will be decreasing due to growing land-use pressures. In some instances and locales stresses may reach a point where recovery of the system becomes so expensive in time, effort, money, and energy as to make it impossible. It is also clear that the vital relationships between the ecological, the social, and the technical systems of our arid and semiarid lands are just beginning to be recognized and understood by decision makers and politicians. At this point of convergence new analytical tools will have to be made available to managers in the public and private sectors and at all levels of government. Risk/benefit analysis is one such tool that may help in formulating a wider range of options, in better evaluating the short- and long-term consequences, in selecting sounder alternatives, and in providing flexibility and responsiveness to the decision-making process.

REFERENCES

Bureau of Indian Affairs, USDI, 1975-79, "Navajo Area, Soil and
    Range Inventory, Technical Reports."
Bureau of Land Management, USDI, 1975, "Range Condition Report"
    prepared for the Senate Committee on Appropriations.
Bureau of Reclamation, USDI, 1975, "Westwide Study Report on Critical
    Water Problems Facing the Eleven Western States."
Fairchild, Warren D., 1973, The role of water in the energy crisis,
    in: "The Role of Water in the Energy Crisis," Karen E. Stork,
    ed., proceedings of a conference, October 23-24, 1973,
    Nebraska Water Resources Research Institute, Lincoln, Nebraska,
    prepared for OWRT/USDI.
Gertsch, W. Darrell, et al., 1977, "Water Requirements for Future
    Energy Development in the West: State Perspectives," ERDA.
Grew, Priscilla, 1979, "California Soils: An Assessment," California
    Department of Conservation, Draft report.
Howard, Arthur D., and Remson, Irwin, 1978, "Geology in Environmental
    Planning," McGraw-Hill, New York.
National Water Commission (NWC), 1973, "Water Policies for the
    Future," Final Report to the President and Congress of the U.S.
Templer, Otis W., April 1978a, Texas ground water law: Inflexible
    institutions and resource realities, Ecumene, 10(1):6.
Templer, Otis W., 1978b, "The Geography of Arid Lands: A Basic
    Bibliography," ICASALS/Texas Tech Univeristy, Lubbock.
Trelease, Frank J., 1976, "The Role of Water Law in Developing the
    American West," International Conference on Water for Peace.

THE VALUE OF A LIFE: WHAT DIFFERENCE DOES IT MAKE?

John D. Graham* and James W. Vaupel**

*National Academy of　　　**Department of Policy Sciences
　　Sciences　　　　　　　　and Business Administration
Washington, D.C.　　　　　Duke University
　　　　　　　　　　　　　Durham, North Carolina

Critics of benefit-cost analyses of lifesaving programs commonly dismiss such analyses with the query "but how can you put a dollar value on a life?" Some believe that it is "morally and intellectually deficient"[1] to attempt to monetize mortality. Other critics have observed that there are, at least currently, no generally agreed upon estimates of the so-called "value of a life" and, consequently, as Nicholas Ashford of M.I.T. has argued, "until society better understands this value, current analytic valuations of life must always be inadequate, and cannot be directly compared with the monetary costs or benefits of a regulation."[2]

That no consensus exists about how to express in dollars the benefits of averting deaths is certainly correct. Although the advocates of "willingness-to-pay" measures have gained the offensive against defenders of the "foregone earnings" (or "human capital") approach, the internecine battle here is by no means done.[3] Within the willingness-to-pay community, a subdued and often unacknowledged debate pits those who value lives against a smaller but persuasive-- group who value life-years; in a second debate, those psychologists and decision analysts who ask individuals their preferences question the methods of the economists who impute safety preferences as revealed by wage premiums for hazardous occupations. Surveys of expressed willingness-to-pay for small reductions in the probability of death have yielded values of a life from $50 thousand to $8 million (in 1978 dollars).[4] Nine recent labor market studies of wage premiums have produced a narrower, but still disparate, range of values spread roughly evenly from $300 thousand to $3.5 million.[5]

233

In researching this article, we scrutinized some 35 studies of
the costs and benefits of health, safety, and environmental programs.
As might be expected, given the disarray both among the theorists
who attempt to define the value of a life and among the empiricists
who attempt to measure it, the practitioners of these policy analyses differed considerably in how they valued lives. Of the 35 studies, 24 were benefit-cost analyses that explicitly assigned dollar
values to lives saved, whereas 11 were cost-effectiveness analyses
that estimated cost per life saved.[6] Of the 24 studies that valued
lives, 15 used a foregone-earnings value, seven used a willingness-
to-pay value, and two used values that were claimed to be consistent
with both the foregone-earnings and the willingness-to-pay approaches.
Five of these analyses used ranges of values; the other 20 picked
point estimates--ranging from $55 thousand to $7 million.

In the seven studies that relied on willingness-to-pay estimates,
the median value of a life was $625 thousand; in the 15 foregone-
earnings studies, it was $217 thousand, only roughly a third as
much.[7] At least to theorists this disparity may be unsettling,
since the foregone-earnings approach has little theoretical support.
Consequently, benefit-cost analyses based on foregone-earnings values
may be undervaluing the benefits of lifesaving programs. Although
15 studies used foregone-earnings values and only seven used willingness-to-pay values, an encouraging trend is that of the ten most
recently published studies, half used a willingness-to-pay measure.

Given the uncertainties about how to define, let alone measure,
the value of a life, it might be expected that the authors of the
benefit-cost studies would calculate--and the reviewers and editors
would demand--the most careful sensitivity analyses of how robust
their conclusions were to alternative assumptions about the monetary
value of lifesaving. Just seven of the 24 benefit-cost studies,
however, contain any sensitivity analysis at all, and only two studies identify the "switch-point" or "breakeven" value that determines
when a policy option should be favored over the contending alternative. Frequently, the estimates of mortality risks used in these
studies are even more uncertain than the value of a life, making
the absence of sensitivity analysis even more inexplicable--and in-
excusable. Beyond this, most of the studies are afflicted with a
variety of sins of omission and commission that we intend to detail
in another paper. Even those who favor analysis in principle have
to admit that analysis in practice is so devilishly demanding that
the most diligent, intelligent, and well-intentioned practitioners
often go astray.

Despite the resultant high level of noise, comparison of the
35 analyses of lifesaving programs does lead to some intriguing, if
broadbrush, conclusions. To facilitate comparisons across studies,
we calculated the "additional cost per additional life saved" of

going from one policy option (usually, but not always, the status quo) to some alternative. Since some of the 35 studies considered several policy alternatives, we were able to compute a cost per life saved for 57 policy pairs. In each instance, we computed a net cost by subtracting from total costs any non-mortality benefits that the authors of the studies estimated: We made no attempt to correct for omitted costs or benefits. A number of analysts have cogently argued that since lives are never saved but merely prolonged, it is also informative to consider cost per life-year saved. Consequently, we estimated this figure for each of the 57 policy pairs as the quotient of the cost per life saved and the average life expectancy gained by individuals whose lives were saved.[8]

Table 1 summarizes the results. A number of interesting patterns and conclusions emerge.

First, for over a quarter of the policy pairs (13 of 57), the net costs are less than zero even when the benefits of saving lives are ignored. These lifesaving programs are justified by various morbidity and non-health gains alone: the mortality reductions achieved can be viewed as a generous bonus.

For many of the remaining policy pairs, the cost per life saved is low. Two judicious students of benefit-cost analysis have surveyed the theoretical and empirical literature to estimate a reasonable range for the value of a life: Martin Bailey's "low" estimate is $170 thousand and Robert S. Smith's plausible lower bound is $300 thousand.[9] For 59% (34 of 57) of the policies pairs in Table 1 the cost per life saved is under Bailey's $170 thousand and for 65% (37 of 57) it is less than or equal to Smith's $300 thousand. Thus, although benefit-cost analysis is sometimes criticized as being biased against health, safety, and environmental policy, for some three-fifths of the policy pairs examined benefit-cost analysis strongly supported lifesaving programs.

Professor Bailey's "high" estimate of the value of a life is $715 thousand, whereas Professor Smith's plausible upper bound is $3 million. In 16 cases, (28%), the cost per life saved exceeds Bailey's value and in 12 cases (21%) it exceeds Smith's. Thus in roughly a quarter of the policy pairs we compared, the additional benefits of a lifesaving program would not appear, at least to a benefit-cost analyst, to be worth the additional costs.

That leaves relatively few cases in the middle. In only seven cases (12%) does the cost per life saved fall within Bailey's range from $170 to $715 thousand, and similarly, in only eight cases does it fall within Smith's order of magnitude range from $300 thousand to $3 million. Furthermore, in only 11 cases does the cost per life save fall within the wide combined range from $170 thousand to $3 million.

TABLE 1. Summary of Results

Problem Area	Agency Concerned	Author	Base Case Policy Option	Alternative Policy Option	Net Additional Cost of Alternative Policy Option	
					Per Life Saved	Per Life-Year Saved
Highway Safety	NHTSA	Warner[15] (1975)	Status quo	Mandatory air bags	$ 0	$ 0
"	"	DOT[17] (1976)	Status quo	Mandatory passive belts	0	0
"	"	COWPS[19] (1977)	Status quo	Compulsory belt usage law	0	0
"	"	Clotfelter[20] & Hahn (1978)	Status quo prior to 55 mph limit	55 mph speed limit	0	0
"	"	NHTSA[21] (1976)	Status quo	Roadside hazard removal	0	0
"	"	"	Status quo	Traffic enforcement	0	0
"	"	"	Status quo	Vehicle inspection	0	0
"	"	Muller[43] (1980)	Voluntary moto-cycle helmet usage	Compulsory helmet usage law	0	0
Genetic Screening	HHS	Swint et al.[22] (1979)	Status quo	Community screening program	0	0
Clothing	CPSC	Dardis et al.[23] (1978)	No law	Clothing flammability law	0	0
Smoke Detectors	CPSC	Watterman et al.[24] (1978)	Status quo	Mandatory smoke detectors	0	0
Stationary Source Air Pollution	EPA	Koshal & Koshal[25] (1973)	Pre-1970 conditions	1970 Clean Air Act Standards	0	0
Stationary Source Air Pollution	EPA	Crocker et al.[26] (1979) & CEQ[27] (1980)	Pre-1970 conditions	1970 Clean Air Act Standards	0	0
Highway Safety	NHTSA	COWPS[19] (1977)	Status quo	Mandatory passive belts	3,600	88
"	"	Zeckhauser & Shephard[29] (1976)	Status quo	Mandatory air bags	13,000	530
Heart Disease Policy	HHS	"	Status quo	Mobile CHD unit	15,000	1,800
Highway Safety	NHTSA	Gates[16] (1975)	Status quo	Active lap/ shoulder belts	21,000	516
Stationary Source Air Pollution	EPA	Lave & Seskin[30] (1977)	Pre-1970 conditions	1970 Clean Air Act Standards	30,000	2,300
Smoke Detectors	CPSC	Potter et al.[31] (1976)	Status quo	Mandatory, in sleep-ing rooms only	40,000	1,300
Highway Safety	NHTSA	Grabowski & Arnould[18] (1980)	Status quo	Mandatory passive belts	40,700	1,000
"	NHTSA	NHTSA[21] (1976)	Status quo	Emerging medical services program	41,000	1,000
Stationary Source Air Pollution	EPA	Freeman[28] (1980) CEQ[27] (1979)	Pre-1970 conditions	1970 Clean Air Act	50,000	3,800
Highway Safety	NHTSA	Zeckhauser & Shephard[29] (1976)	Status quo prior to 55 mph limit	55 mph limit with full adherence	59,000	2,500
Furniture Fires	CPSC	SRI[32]	Status quo	Mandatory smoke detectors	60,000	1,900
Highway Safety	NHTSA	Zeckhauser & Shephard[29] (1976)	Status quo prior to 55 mph limit	55 mph limit with partial adherence	64,000	1,900
"	"	DOT[17] (1976)	Status quo	Mandatory air bags	78,000	1,900
"	"	NHTSA[21] (1976)	Status quo	Alcohol safety action projects	81,500	2,000
"	"	COWPS[19] (1977)	Status quo	Mandatory air bags	94,000	2,300

TABLE 1.   Summary of Results (continued)

Problem Area	Agency Concerned	Author	Base Case Policy Option	Alternative Policy Option	Net Additional Cost of Alternative Policy Option	
					Per Life Saved	Per Life-Year Saved
Highway Safety	NHTSA	DOT[17] (1976)	No restraint	Active lap/shoulder belt system	94,000	2,300
Heart Disease Policy	HHS	Zeckhauser & Shephard[29] (1976)	Status quo	Diet program	102,000	6,500
Highway Safety	NHTSA	Robertson[34] (1977)	Status quo	Mandatory air bags	117,000	2,800
Saccharin	HHS	COWPS[35] (1977)	Status quo	Ban	136,000	8,500
Highway Safety	NHTSA	Zeckhauser & Shephard[29] (1971)	Mandatory air bags	Mandatory air bags plus 55 mph limit with full adherence	148,000	6,000
"	"	Gates[16] (1975)	No restraint	Mandatory air bags with active lap belts	162,000	4,000
"	"	GAO[33] (1976)	Pre-1966 conditions	1966 Motor Vehicle Safety Act	255,000	6,300
"	"	Gates[16] (1975)	Status quo	Mandatory air bags	300,000	7,300
Pertussis Vaccine	HHS	Kaplan et al.[36] (1979)	Immunize	No program	300,000	4,200
Furniture Fires	CPSA	SRI[32] (1979)	Mandatory smoke detectors	CPSC flammability standards	400,000	12,900
Highway Safety	NHTSA	Grabowski & Arnould[18] (1980)	Status quo	Mandatory air bags	408,000	10,000
"	"	Castle[37] (1976)	65 mph limit	55 mph limit	500,000	12,000
"	"	Ford[38] (1979)	Unsafe fuel tank	Safer fuel tank	686,000	17,000
Smoke Detectors	CPSC	Potler et al.[31] (1979)	Mandatory, in sleeping rooms only	Mandatory in all rooms	1,000,000	32,000
Highway Safety	NHTSA	C. Lave[39] (1979)	Status quo prior to 55 mph speed limit	55 mph speed limit	1,200,000	29,000
Mobile Source Air Pollution	EPA	NAS[40] (1974)	Pre-1970 conditions	1970 Clean Air Act	1,350,000	105,000
Highway Safety	NHTSA	Arnould & Grabowski[18] (1980)	Mandatory passive belts	Mandatory passive belts and air bags	1,400,000	34,000
Acrylonitrile	OSHA	COWPS[41] (1978)	Status quo	2.0 ppm	3,520,000	230,000
Carcinogens in water	EPA	COWPS[42] (1978)	150 mcl rule	100 mcl rule	3,800,000	240,000
"	"	COWPS[42] (1978)	Status quo	150 mcl rule	3,900,000	240,000
Arsenic	OSHA	COWPS[44] (1978)	5 mcl rule	.004 mcl rule	5,000,000	390,000
Carcinogens in Water	EPA	COWPS[42] (1978)	100 mcl rule	50 mcl rule	6,300,000	390,000
Vinyl Chloride	OSHA	Perry & Outlaw[45] (1978)	50 ppm	1 ppm	7,500,000	490,000
Benzene Emissions	EPA	Nichols[46] (1980)	No control	97% control	7,600,000	480,000
Coke ovens	OSHA	COWPS[47] (1976)	Status quo	Proposed OSHA standard	12,100,000	790,000
Acrylonitrile	"	COWPS[41] (1978)	2.0 ppm	1.0 ppm	28,800,000	1,900,000
Benzene Emissions	EPA	Nichols[46] (1980)	97% control	99% control	51,000,000	3,200,000
Benzene	OSHA	Wilson[48] (1979)	10 ppm rule	1 ppm rule	102,000,000	6,600,000
Acrylonitrile	"	COWPS[41]	1.0 ppm	0.2 ppm	169,200,000	11,000,000

This is an encouraging finding since it implies that the specific value of a life used in a benefit-cost analysis has not, in something like four-fifths or five-sixths of the cases, altered the
policy implications of the study.[10] Given the confusion in the
theory and practice of valuing lives, it is reassuring that precise
estimates of the value of a life were usually not needed. In a
prescient observation made prior to the recent spate of benefit-
cost studies of lifesaving programs, Richard Zeckhauser argued:

> there are conceptual and philosophical difficulties inher-
> ent in any procedure that attempts to attach a value of
> life, though conducting assessments with the aid of such
> procedures may nevertheless be helpful. In many circum-
> stances policy choices may not change substantially if
> estimates of the value of life vary by a factor of ten.
> Getting a valuation that is accurate within a factor of
> three might be very useful.[11]

Our results support Zeckhauser's optimism.

Beyond this, the results suggest that it is usually not necessary to explicitly value lives: Instead of a benefit-cost analysis,
a cost-effectiveness analysis that calculates cost per life saved
may often be sufficient. Given the controversy about assigning
monetary values to the benefits of saving lives--and the distaste-
fulness to many of doing so--it would seem to be judicious to rely
on cost-effectiveness analysis to the extent possible.[12] Indeed,
decision makers and other readers of these studies may be at least
as interested in knowing that the cost per life saved by some program is $10 thousand--or $10 million--as in knowing that estimated
net benefits amount to -$35 million or that the estimated benefit/
cost ratio is 1.7.[13]

Table 2 cross-tabulated the 57 policy pairs in Table 1 by the
agency concerned and by three ranges of cost per life saved. Since
the studies we surveyed may not be representative and since they
suffer from a myriad of empirical and theoretical flaws, implications should be drawn from Table 2 with caution. Nonetheless, the
table does suggest that the costs of saving lives differ greatly
across agencies or, at least, that the policy options being weighed
by different agencies vary considerably in cost-effectiveness.

Another rough indication of inter-agency disparities is given
by the median values of the cost per life saved for each agency's
range of policy options. For the National Highway Traffic Safety
Administration (NHTSA), the Department of Health and Human Services
(HHS), and the Consumer Produce Safety Commission (CPSC), the medians
are comparable: $64 thousand, $102 thousand, and $50 thousand,
respectively. For the Environmental Protection Agency (EPA), however,

Table 2.   Breakdown of Policy Options by Agency
and by Net Cost Per Life Saved

Number of Cases Where Net Cost
Per Life Saved Is:

Agency	Under $170,000	Between $170,000 and $3,000,000	Above $3,000,000	Total
NHTSA	22	7	0	29
HHS	4	1	0	5
CPSC	4	2	0	6
EPA	4	1	5	10
OSHA	0	0	7	7
Total	34	11	12	57

the median is $2.6 million . . . and for the Occupational Safety
and Health Administration (OSHA) it is $12.1 million.

In addition to data on cost per life saved, Table 1 also presents estimates of cost per life-year saved. Those policies that
are most cost-effective in saving lives also tend to be the policies
that, by preventing accidents and acute diseases, save the lives of
younger individuals. On the other hand, those policies that are
least cost-effective in saving lives tend to be the policies that
focus on preventing various kinds of cancer and chronic disease
that largely afflict the elderly. For example, the victims of
motor vehicle accidents lose, on average, 41 years of life expectancy, whereas the victims of cancer lose 16 years.[14] Consequently,
measuring performance in cost per life-years saved does not substantially alter the rank order of the programs. Indeed, the Spearman rank correlation coefficient, for the 44 policy pairs with a
positive value of cost per life saved, is 0.98. The debate between
the advocates of cost per life-year saved versus cost per life saved
thus may be more of theoretical interest than of operational significance, at least in setting priorities.

Measuring performance in cost per life-year saved does, however,
further widen the large differences among the various types of lifesaving programs. The least expensive OSHA program is seven times
more expensive per life-year saved than the most expensive NHTSA program; the median OSHA program is more than 400 times more expensive
per life-year saved than the median NHTSA program. Again, these
findings should not be taken as anything more than suggestive, since

'they are based on a crude comparison of a set of disparate studies.
Furthermore, policymakers, for numerous legitimate reasons, may
explicitly decide to devote more resources to saving lives in some
areas than in others. For example, some causes of death are particularly painful and anxiety producing. To give another example,
some causes of death may seem especially "unfair" since they result
from largely involuntary exposure to, say, carcinogens in the air
rather than from more voluntary factors such as cigarette smoking.
The question, however, remains: how _much_ more is it reasonable to
spend in some areas than in others? Do the huge disparities in
lifesaving expenditures reflect defensible judgments? To pose the
question differently, could society, by shifting resources into
more cost-effective lifesaving programs, save enough additional
lives to justify such a shift?

The striking discrepancies across agencies and programs in
cost per life saved, and the even greater discrepancies in cost per
life-year saved, that our admittedly roughhewn study has uncovered
suggest that more careful, larger scale efforts at comparing opportunities for saving lives across agencies and policy options may
constructively contribute to the political process of setting
health, safety, and environmental priorities. The confusion about
the value of a life does not imply that thoughtful quantitative
analysis cannot help us sort out our confusion about how best to
save lives.

REFERENCES

1. Michael S. Baram, "Regulation of Health, Safety and Environ-
      mental Quality and the Use of Cost-Benefit Analysis," final
      report to the Administrative Conference of the United States
      (March 1, 1979) p. 27.
2. "Benefits of Environmental, Health, and Safety Regulation,"
      prepared for the Senate Government Affairs Committee,
      Center for Policy Alternatives, MIT (March 26, 1980) p. 19.
3. The "human capital" measure is based on estimates of the pres-
      ent value of foregone earnings due to premature death. The
      "willingness to pay" measure is derived from estimates of how
      much individuals are willing to pay to reduce their probabil-
      ity of death by a small amount.
4. For the lower bound, see Jan P. Acton, "Evaluating Public Pro-
      grams to Save Lives: The Cost of Heart Attacks," Rand Corp.,
      Santa Monica, Cal. (1973); for the upper bound, see M. W.
      Jones-Lee, "The Value of Life," University of Chicago Press,
      Chicago (1976).
5. For a review of these studies, see Robert S. Smith, Compensat-
      ing wage differentials and public policy: a review, Ind. &
      Labor Relations Rev., 32, No. 3 (1979), pp. 339-52.

6. All six of the studies done by the Council of Wage and Price Stability were cost-effectiveness studies.

7. The mean value of a life in the willingness-to-pay studies was $1,288 thousand; in the foregone-earnings studies it was $204 thousand.

8. Let p_i be the proportion of those individuals whose lives would be saved who are age i and let e_i be the life expectancy of individuals age i. Then "average life expectancy gained" is given by the sum over all ages i of the product of p_i and e_i. We used life expectancy data for the U.S. population for 1976 as given in the "Monthly Vital Statistics Report," 26, No. 11 (Feb. 1978).

9. Martin Bailey, "Reducing Risks to Life: Measurement of the Benefits," American Enterprise Institute, Washington, D.C. (1980), pp. 52-66; Robert S. Smith, op cit.

10. Of course, if a study considers a continuous range of alternatives rather than a few discrete alternatives, the value of a life will influence which policy is optimal.

11. Richard Zeckhauser, Procedures for valuing lives, Public Policy, 23, No. 4 (1975), pp. 419-464.

12. For further discussion of this, see Howard Raiffa, William Schwartz, and Milton Weinstein, Evaluating health effects of social decision and programs, "EPA Decision Making," National Academy of Sciences, Washington, D.C. (1978).

13. In most cases, all three kinds of statistics should be presented to provide a variety of perspectives.

14. These life-expectancy statistics are based on the assumption that victims, if saved, would face the same life chances as non-victims. Victims, however, may be frailer or more accident prone, on average, than non-victims. By making some estimates about this, Richard Zeckhauser and Donald Shephard in Where now for saving lives, Law and Contemp. Problems, 40, No. 4 (Autumn 1976), estimate an average life expectancy gained of 25 years, rather than 41 years, for victims of motor vehicle accidents. Also see James W. Vaupel, Kenneth G. Manton, and Eric Stallard, The impact of heterogeneity in individual frailty on the dynamics of mortality, Demography, 16, No. 3 (August 1979).

15. Charles Y. Warner, Michael R. Wither, and Richard Peterson, Societal priorities in occupant crash protection, Fourth International Congress on Automotive Safety (July 1975), pp. 907-960.

16. Howard P. Gates, Jr., Review and critique of NHTSA's revised restraint system cost-benefit analysis, 4th Int. Congress on Automotive Safety (July 1975), pp. 209-233.

17. William T. Coleman, Secretary of the Department of Transportation, Benefit-cost analysis of motor vehicle occupant crash protection, Federal Register (June 14, 1976), pp. 24078-79.

18. Richard J. Arnould and Henry Grabowski, Auto safety regulation: an analysis of market failure, Bell J. Econ. & Management Sci., in press.

19. Council on Wage and Price Stability, Occupant Crash Protection Standard, Executive Office of the President, Washington, D.C., CWPS-244:1-24 (May 31, 1977).

20. Charles T. Clofelter and John C. Hahn, Assessing the national 55 mph speed limit, Policy Sci., 9:281-294 (1978).

21. "National Highway Safety Needs Report," U.S. Department of Transportation, Washington, D.C. (April 1976).

22. Swint, Shapiro, Corson, Reynolds, Thomas, and Kazazian, The economic returns to community and hospital screening programs for genetic disease, Preventive Medicine, 8:463-470 (1979).

23. Rachel Dardis, Susan Aaronson and Ying-Nan Lin, Cost-benefit analysis of flammability standards, Am. J. Ag. Econ., pp. 697-699+ (November 1978).

24. T. E. Waterman, K. R. Minszewski and D. G. Spandoni, "Cost-Benefit Analysis of Fire Detectors," 63 pp., IIT Research Institute, Chicago (September 1978).

25. Rajindar Koshal and Manjulika Koshal, Environments and urban mortaility--an econometric approach, Environmental Pollution, 4:247-259 (1973).

26. Thomas D. Crocker, William D. Schulze, Shaul Ben-David, and Allen V. Kneese, "Methods Development for Assessing Air Pollution Control Benefits," vol. 1, Experiments in the Economics of Air Pollution Epidemiology, prepared for EPA (February 1979).

27. Council on Environmental Quality, Annual Report, Washington, D.C. (1980). This report estimates incremental costs of stationary source cleanup at $7 billion.

28. A. Myrick Freeman, III,"The Benefits of Air and Water Pollution Control: A Review and Synthesis of Recent Estimates," prepared for the Council on Environmental Quality (December 1979).

29. Zeckhauser and Shepard, op. cit.

30. Lester B. Lave and Eugene Seskin, "Air Pollution and Human Health, Resources for the Future," Johns Hopkins University Press (1977), Chap. 10, The benefits and costs of air pollution abatement, pp. 209-234.

31. J. M. Potter, M. L. Smith, and S. S. Lanwalker, "Cost-effectiveness of Residential Fire Detector Systems," Texas Tech. University (November 1976).

32. SRI, "Decision Analysis of Strategies for Reducing Upholstered Furniture Fire Losses," Department of Commerce (June 1979).

33. General Accounting Office, "Effectiveness, Benefits and Costs of Federal Safety Standards for Protection of Passenger Car Occupants," Washington, D.C. (July 1976).

34. Leon S. Robertson, Car crashes: perceived vulnerability and willingness to pay for crash protection, J. Community Health, 3, No. 2 (1977), pp. 136-141.

35. Council on Wage and Price Stability, Council Urges More Study
     on Saccharin Ban, Executive Office of the President (June 15,
     1977), p. 1-16.
36. Koplan, Schoerbaum, Weinstein and Fraser, Pertussis vaccine:
     an analysis of benefits, risks, and costs, New England J.
     Med., 301 (Oct. 1979). "No program" is desirable at high
     values of a life because more deaths will be caused by reac-
     tions to a vaccine than will be prevented.
37. Gilbert Castle, The 55 mph limit: A cost-benefit analysis,
     Traffic Eng. (January 1976) pp. 11-14.
38. Ford Motor Co., Benefits and costs related to fuel leakage asso-
     ciated with the Static Rollover Test portion of FMVSS 208,
     reprinted in the Chicago Tribune (October 14, 1979), section
     1, p. 1.
39. Charles A. Lave, Energy policy as public policy, in: "Changing
     Energy Use Futures," vol. 4, pp. 2046-53, Fazzolare and Smith,
     eds., 2nd Int. Conf. on Energy Use Management (October 22-26,
     1979).
40. National Academy of Sciences, "Air Quality and Automobile Emis-
     sion Control," vol. 4, pp. 1-471, The Costs and Benefits of
     Automobile Emission Control, prepared for U.S. Senate Commit-
     tee on Public Works (September 1974).
41. Council on Wage and Price Stability, Council Comments on Pro-
     posed Standard for Occupational Exposure to Acrylonitrile,
     Executive Office of the President (May 22, 1978).
42. Council on Wage and Price Stability, Comments Submitted to the
     EPA on the Proposed Drinking Water Regulations, Executive
     Office of the President, Washington, D.C. (September 5, 1978).
43. Andreis Muller, Evaluation of the costs and benefits of motor-
     cycle helmet law, Am. J. Public Health, 70, No. 6 (June 1980),
     pp. 586-592.
44. Council on Wage and Price Stability, Council Comments on OSHA's
     Proposed Standard on Arsenic, Executive Office of the Presi-
     dent (September 14, 1976).
45. Charles Perry and Randall Outlaw, "Safe and Healthful Working
     Conditions--The Vinyl Chloride Experience," University of
     Pennsylvania, Philadelphia (1978).
46. Albert Nichols, Alternative regulatory strategies for control-
     ling benzene emissions from malcic anhydride plants, paper
     for the Environmental Protection Agency (March 1980).
47. Council on Wage and Price Stability, Exposure to Coke Oven
     Emissions: Proposed Standard, Executive Office of the Presi-
     dent (May 11, 1976).
48. The Supreme Court's Benzene Decision, Secretary of Labor vs.
     API (July 2, 1980). The Court cites Richard Wilson's work
     suggesting that the 1 ppm benzene standard would avert only
     2 cancer deaths every six years. Ignoring capital costs and
     using OSHA's estimate of $34 million per year in operating
     costs, it appears that the 1 ppm standard would cost $102
     million per life saved.

ON THE VALUE DEPENDENT ROLE OF THE IDENTIFICATION PROCESSING AND EVALUATION OF INFORMATION IN RISK/BENEFIT ANALYSIS

Andrew P. Sage and Elbert B. White

Department of Engineering Science and Systems
University of Virginia
Charlottesville, Virginia

INTRODUCTION

The decision-making process may be described as the identification and evaluation of a number of alternatives potentially capable of resolving an issue, and the ultimate selection and implementation of one (or more) of the several available alternative actions so as to satisfy, to the maximum extent possible, the goals or objectives of the decision maker [Efstathiou and Rajkovic, 1979]. The decision is one of "certainty" if the decision maker has complete and accurate knowledge of the consequences that will follow from each alternative. The decision is one under "risk" if there is accurate knowledge of the probability distribution of consequences of the alternatives. Finally, decisions occur under "uncertainty" if the decision maker cannot assign any definite probabilities to outcomes which follow from decision alternatives. When the alternatives are many, when there are risks or uncertainties about the consequences of the different actions, and when there are multiple noncommensurable conflicting objectives which result, the decision process is very complicated. This is typically the case in contemporary issues regarding risk and hazards.

Societal choicemaking, in the public and private sectors, is being studied by researchers from an increasingly diverse set of disciplines including medicine, economics, education, political science, geography, systems engineering, marketing, management science, and psychology [Hogarth, 1975; MacCrimmon and Larson, 1979; Milburn and Billings, 1976; Okrent, 1975; Sage, 1981a; Slovic, Fischhoff, and Lichtenstein, 1977; Tversky, 1969]. Certainty is rarely encountered in actual choicemaking situations. Lack of full control over the event outcomes that result from implementation

of particular action options means that risk is encountered. Lack
of information about the likelihood of occurrence of outcome events
in risky situations brings about conditions of uncertainty. The
growing difficulties of regulation, setting of standards, legislation, environmental protection, economic efficiency, technological
choice, and individual liberty necessitate better understanding of
risk and improved methods for integrating societal and technical
judgment on issues involving risk [Greenberg, Marshall, and Yawitz,
1978; Roberts, 1978; Swieringa et al., 1976].

    One of our principal roles as systems analysts is to assist
clients in the formulation and consistent expression of facts and
values and the proper aggregation of these to enable selection of
policies. More specifically, our task is to help decision makers
identify issues and quantify variables that are especially important to the resolution of complex issues. This paper discusses
cognitive heuristics, cognitive biases, and value inconsistencies
as well as their effects on decision-makers' evaluation processes
and the extent to which debiasing activities may be implemented by
normative models of decision making. It compliments a recently
published detailed survey paper [Sage and White, 1980].

DESCRIPTIVE DECISION BEHAVIOR

    Since Bernoulli (1738), the standard criterion for rational
decision making in environments involving risk and uncertainty has
been the maximization of expected utility. Important subsequent
developments in the field were the axiomatization of utility theory
by von Neumann and Morgenstern [1947] and the incorporation of subjective probability by Savage [1954]. The notion of a subjectively
expected utility (SEU) model was also advanced by Edwards [1954].
Keeney and Raiffa [1976], Sage [1977], and others have described the
standard approach used for the assessment of multi-attribute von
Neumann and Morgenstern utility functions.

    Although subjective expected utility theory originated as a
normative model of human behavior, it has also been proposed and
tested as a descriptive model. Many studies in the behavioral literature indicate that subjective expected utility theory is, at
times, a poor descriptor of empirically observed decision-making
behavior. This is typically due to the use of "boundedly rational"
processes, which may include cognitive bias in information processing and use of poor cognitive heuristics rather than substantively
rational decision rules [Karmarkar, 1978; Sage and White, 1980].
There are, of course, a number of studies which indicate relatively
good performance of substantive decision behavior with the subjective expected utility theory [Okrent, 1975; Roberts, 1978]. Virtually no studies, however, indicate that unaided (naive) subjects

will, in practice, use the "process" of SEU theory, even though substantive results of judgment may be those of the SEU theory. Thus,
an important issue is the determination of when actual behavior will
deviate from substantive rationality and the provision, through
appropriate aids, of corrective procedures.

Cognitive psychologists generally report that, in many situations, subjective expected utility models do not describe the
descriptive decision behavior observed in individuals [Allais and
Hagen, 1979; Goldberg, 1976; Kahneman and Tversky, 1979; Koriat,
Lichtenstein, and Fischhoff, 1980; Poulton, 1977; Sjoberg, 1979;
Skweder, 1980]. The postulate or axiom which subjects most often
violate is the substitution principle, or the equivalent axiom of
von Neumann and Morgenstern [Allais, 1953; Coombs, 1975; Coombs and
Huang, 1976; MacCrimmon and Larson, 1979]. Also, individuals do
not often process probability information as they should, and value
inconsistencies are often noted as well [Fischhoff, Slovic, and
Lichtenstein, 1980; Slovic and Lichtenstein, 1968].

The 1953 Allais Paradox [Allais, 1953] was one of the earliest
demonstrations of some inconsistencies surrounding the use of the
then-existing SEU theory as a descriptive model. The paradox,
included in an extensive questionnaire which Allais circulated to
a number of prominent economists, asked subjects to choose between
the lotteries $A(t)$ and $B(t)$, where:

$A(t)$ gives: \$1 million with probability $1 - t$

$\quad\quad\quad\quad$ \$0 million with probability t

$B(t)$ gives: \$5 million with probability 0.10

$\quad\quad\quad\quad$ \$1 million with probability $0.89 - t$

$\quad\quad\quad\quad$ \$0 million with probablity $0.01 + t$

It was observed that people typically prefer $A(0)$ to $B(0)$
and $B(0.89)$ to $A(0.89)$. This is inconsistent with the simplest
single-attribute formulation of subjective expected utility theory,
which requires that, if $A(0)$ is preferred to $B(0),(A(0) > B(0))$,
then $A(t)$ is preferred to $B(t),(A(t) > B(t))$ for all $0 < t <$
0.89.

Kahneman and Tversky [1979] describe some effects, such as
certainty, reflection, and isolation, which lead unaided experimental subjects to violate the von Neumann and Morgenstern axioms
of normative decision theory. The certainty effect is one in which
people often overweigh outcomes considered certain relative to
those that are merely very probable. The reflection effect is

where people reverse their preference order, compared to that predicted from the certainty effect, when the momentary amounts of a
prospect is reflected about zero. Thus certainty increases the
desirability of gains and increases the repugnance of losses as
well. This is incompatible with the notion, often expressed concerning risks in business situations, that certainty is generally desirable. It appears desirable only when the outcomes are beneficial.
The isolation effect results when people disregard components common
to all alternative outcomes or prospects and focus only on incremental
components that distinguish them. Thus, value or utility is determined by changes of wealth rather than final asset position, which
includes current wealth. The isolation effect may produce inconsistent preferences since it is possible to decompose prospects in several ways. Sage and White [1980] present a detailed survey of these
and other observed departures from subjective expected utility
theory.

These failings have almost always been attributed to cognitive
limitations associated with predecisional activities. For example,
the difficulty of taking into account all salient properties of compared or estimated alternatives forces people to employ simplifying
strategies or cognitive "heuristics" instead of the aggregation of
facts and values prescribed by SEU theory.

Cognitive biases can be described as those influences which
cause one's beliefs to not properly reflect the state of information [Einhorn and Hogarth, 1981; Fischhoff, 1975]. Prevalent among
the cognitive biases identified as the use of "availability," where
the probability estimate is based on what the estimator recalls or
can visualize; "representativeness," where the probability estimate
of an event is based on the degree to which it is representative of
or similar to a major characteristic of the population from which
it is drawn; and "anchoring," where the estimate is affected by
an initial value from which one adjusts in order to yield a final
answer. These identifiable simplifying information-processing
strategies result in cognitive bias which often leads to "social
traps," the result of which is inappropriate judgment.

Among explanations traditionally offered for observed "viola-
tions" were that the subject was bored, tired, disinterested, or
careless, and that the apparent inconsistencies reflect only nominal
"error variance" [MacCrimmon and Larson, 1979] associated with this.
However, it is now commonly accepted that such charges are not fully
correct and much more effort needs to be exerted to resolve information processing deficiencies and thereby improve human judgment by
appropriate corrective procedures.

An explanation often offered for the certainty effect, for
example, is that, when one alternative provides a sure chance of

obtaining a very desirable consequence, people usually select it,
even if it entails passing up a much larger amount that results with
a probability of slightly less than one. When, however, the chances
of winning are small and close together, people usually take the
option that provides the larger payoff. Thus, the decision rule
changes as a function of the contingency task structure, or problem
description. This suggests that there exists an attribute--regret--
that exists in addition to the conventionally assumed single,
monetary-value attribute.

The classical subjective expected utility criterion involves
the multiplication of probability by a utility-of-payoff term. It
is assumed that there exists independence between probabilities and
payoffs. One consequence of this independence is the inability of
switching from a conservative style to one of trying to win big
[MacCrimmon and Larson, 1979]. When the probability levels decrease
or when the payoff levels decrease, then there is a reduced tendency
to pick the alternative giving the lower payoff with the higher probability. When the probability levels decrease, the rationale is one
of viewing the probability difference as insignificant and thus
"going for broke" on the larger payoff, since we often assume that
winning is unlikely in all cases and we may as well strive for the
biggest gain. When the payoff levels decrease, the rationale is one
of "going for broke" since the amount one is able to get for such
does not mean that much in terms of life-time security, etc. It has
been shown that people are quite alert to changes in payoff and probability and hence choose differentially. Thus it is apparent that
the particular parameter values themselves play a major role in
whether one violates the utility independence conditions [MacCrimmon
and Larson, 1979]. Again, we see the presence of lexicographic decision rules: rules which may be extraordinarily poor or reasonably
good depending upon the problem at hand and the heuristic used.

Examples of humans perceiving what they wish, as well as what
they expect to perceive, are readily available. In fact, any current intense emotional state may influence judgment and hence is
cause for concern. Fear of making a mistake may cause an unconscious
"play it safe" bias, and personal identification with any aspect of
the situation may lead to irrational influences [Thaler, 1980].

All individuals are products of their culture. They have powerful perscriptive and proscriptive values and information-processing
patterns imprinted on them. Decision makers, systems analysts, and
engineers should be sensitive to possible emotional or cultural
associations of experts who provide "facts" in the form of risk and
hazard information, for these associations can cause significant
distortions of perceptions, bias in reasoning, and selective block-
age or exaggeration of recall [Albert, 1977; Kahneman and Tversky,
1979].

Although people may not appear to be very risk averse in terms
of their elicited utility function, they may behave in a much more
risk-averse or risk-prone manner in particular circumstances [Kar-
markar, 1979]. Engineering technical studies generally concentrate
on probability distributions of events and on the possible consequences of these, with the hope that the expected value of the distribution is a good parameter for use in policy analysis [Kozielecki,
1975; Morris, 1974]. However, this may not necessarily be the case.
The public contention that an accident with a probability of

10^{-7} and 1 million casualities is not equivalent to an accident with

probability of 10^{-2} and 10 casualties seems to be eminently reasonable. This is especially true in the case of nuclear power plants,
which carry with them the possibility of accidents with large consequences and small probabilities [Morris, 1974]. Of course, this
anomaly can be successfully explained by a risk-aversion coefficient,
as is standard practice in operations research, management science,
and systems engineering. But use of risk aversion and other subjective value-laden constructs is new to most of conventional engineering practice. Use of it in the case presented here easily leads to
the reasonable conclusion that the statistical average value of
life is greater for low-probability catastrophic occurrences than
for higher-probability accidents with lower expected loss of life.

Among other explanations that have been offered for observed
departures from normative SEU theory is the fact that individuals
may ignore or misuse probability information because they may simply
not notice, or not care about, small or infrequent decrements in
reward that result from their ignorance or misuse of the substantively rational theory [Libby, 1976]. Alternatively, they may ignore
or misuse probability information because the time or effort required
to use it properly may be more costly than any anticipated decrease
in payoffs associated with their occasional suboptimal choices
[Libby, 1976].

It is generally recognized that an individual's behavior is in
large part shaped by the manner in which the physical and social
environments are perceived, diagnosed, and evaluated. Similarly,
it is recognized that, in order to experience and cope with the complex, confusing reality of the environment, unaided indivudals have
to form simplified, structured beliefs about the nature of their
world [De Waele, 1978; Hammond, McClelland, and Mumpower, 1980].

There is much empirical evidence to show that decision makers
tend to form heuristic representations for the structure of the
decision rule by conceiving of the risky task in terms of a hierarchical list of dimensions [Kozielecki, 1975]. This kind of con-
ceptualization reduces cognitive strain, although it detracts from
the effectiveness of the final decision.

Also, the order in which a decision maker seeks and evaluates the information of a decision problem is related to the cognitive process leading to final decision [Svenson, 1979]. People may see anticipated relationships that are not even there [Svenson, 1979]. Simplifying strategies may not only be developed from the decision-maker's own experience with similar decision problems, but may also be derived from other people's experiences [Svenson, 1979]. Finally, heuristic decision rule usage may be as much a function of cognitive limits associated with postdecisional activities, such as judgment consequences, as it is a function of those associated with predecision activities, such as assessing probabilities [Thorngate, 1980].

IMPLICATIONS OF COGNITIVE HEURISTICS, INFORMATION PROCESSING, BIAS, AND INCONSISTENT VALUES FOR RISK/BENEFIT ANALYSIS

Two aspects of knowledge important for risk/benefit analysis are what a person believes to be true and the confidence expressed in that belief. It is often possible to assess the correctness of beliefs by looking up information in reference sources. However, evaluating the validity of degrees of confidence is much more difficult [Fischhoff, Slovic, and Lichtenstein, 1980; Sjoberg, 1979]. Only statements of certainty can be evaluated according to whether the beliefs to which they are attached are absolutely true or false [Fischhoff, Slovic, and Lichtenstein, 1980].

Motivational biases occur in cases where assessments do not reflect a persons' conscious beliefs, perhaps because of wishful thinking, self deception, or the distortion of judgment by payoffs and penalties [Einhorn and Hogarth, 1981]. A person's motives are often adaptive, by self regulation, to accommodate their own cognitive limits [Simon, 1957; Thorngate, 1980]. Related adaptations may also occur with respect to the tolerance of, or insensitivity to, decision errors.

Based on these assertions, researchers in risk/benefit analysis concerned with judgment and decision-making processes and those who design adjuvants to assist in planning and decision support must be concerned with understanding the diverse relationships between motives, tolerance, adaptations, stress, cognitive style, and cognitive bias and decision heuristics [Simon, 1979b]. It may well be possible that concepts of aspiration [Lewin et al., 1974], motivation [Messick and McClintock, 1968], stress [Janis and Mann, 1977], adaptation [Helson, 1964], and the contingency task structure of the environment, for example, may be vitally important to the development of information systems adjuvants and other aids to risk/benefit analysis as well as concepts of cognitive limits. It has been demonstrated, for example, that the manner in which the formulation and analysis tasks which preceed evaluation are performed

possibly establishes the boundaries within which the decision is
made [Holsti, 1976]. This emphasizes the need for process studies
of behavior as well as substantive studies.

Research aimed at determining factors underlying the conceptual-
ization of risky tasks has also been conducted [Kozielecki, 1975].
Among the choice conditions examined were those requiring the selective response of "accept-reject." Cognitive styles and personality
traits were found among the factors influential in the development
of heuristic representations of choice situations. Many contemporary researchers strongly support the position that the decision
maker's information-processing and information-evaluation capabil-
ities depend essentially on the task being dealt with, the cognitive
style of the decision maker, and the environment in which the task
occurs [Daniel, 1979; Fishburn, 1974; Kozielecki, 1974 and 1977;
Larichev, Boichenko, and Moshkovich, 1980; Matusewicz, 1975;
Skweder, 1980]. The term contingency task structures is used to
describe the interaction of the environment, the cognitive style
of the decision maker, and the issue at hand. It is this contingency task structure that, in effect, determines the performance
objectives for problem-solving efforts and decision rule selection
[Sage, 1981a and 1981b].

Formulation of appropriate decision-situation structural models
is especially important in risk/benefit issues, for, if an appropriate decision model is not available, the goal of maximizing expected
value may be perceived to be less important than other goals [Thorn-
gate, 1980]. All of this illustrates the necessity of understanding
a decision-maker's motives before prescribing, describing, or explain-
ing the use of various decision heuristics. And, it gives maximum
encouragement to procedurally rational decision-situation models
which reflect, accurately, the objectives of the decision maker. It
is the contingency task structure which determines the performance
objectives for the issue at hand as well as the decision rule
selected for evaluation [Sage and White, 1980; Sage, 1981a].

Under certain preference-elicitation techniques, people may
resort to simplifying strategies such as successive elimination of
alternatives on the basis of various criteria [Larichev, Boichenko,
and Moshkovich, 1980; Svenson, 1979]. In addition, during a decision process, the representation system, or decision-situation structural model, and the sequence of rules applied may be continuously
influencing each other [Svenson, 1979]. For instance, it may be
assumed that a decision maker may change the degree of complexity
of the representation system to meet the requirements of a decision
rule that is felt to be applicable. Correspondingly, the choice of
a decision rule or sequence of rules may be a function of the situation model representing the decision problem, because some of the
simpler rules are only applicable, or make sense, for a limited

number of situation models. For instance, the rule of dominance, which states that the chosen alternative should be better than the other alternatives on at least one attribute and better or equal on all others, works only in a limited number of decision situations since there will generally be more than one nondominated alternative. If the representation of the situation makes it impossible to use the rule of dominance to select the best alternative, then another decision rule must be applied to find the best alternative. Thus, the representation system and the selection of decision rules may affect each other in decision processes. To this end there exists much interest in aggregation rules based on combinations of multiple-objective evaluation and multiple-attribute utility for judgment and choicemaking [White and Sage, 1980].

RESEARCH NEEDS AND RISK/BENEFIT ANALYSIS

If assessments of uncertainty are obscured by biases due to improper information processing devices associated with thinking intuitively about uncertainty, then such assessments will likely not be very useful. In fact, there then exists the need to seek alternatives to the assessment of uncertainties for planning purposes [Einhorn and Hogarth, 1981]. Since each decision evaluation heuristic ignores or misuses some relevant information, heuristics will differ in their relative ability to prioritize alternatives and to yield the optimal choice. Also, since each decision heuristic ignores different features of relevant information, we see that the effects of different information-processing biases will depend upon the heuristic used to evaluate the information [Thorngate, 1980]. The danger with such approaches is that a suboptimal solution may appear to be "optimal" by an inadequate selection of constraints, as has been noted [Milburn and Billings, 1976].

In the short run, use of unrecognized cognitive biases may appear to lead to substantial benefits such as time savings. However, their use can seriously cloud our ability to judge the past so that we may forecast, assess, and plan for the future. We may delude ourselves or others to believe, for example, that we possess hindsightful foresight. However, if the presence of these cognitive biases is not recognized and corrected, then our ability to judge the past or learn from it may be seriously impaired as we become masters of the art of self-deception [Fischhoff, 1975 and 1980; Nisbett and Ross, 1980; Sniezek, 1980].

Almost inevitably, effects introduced by the systems engineering team will bias quantitative subjective assessments, unless very special precautions are taken to prevent, eliminate, or identify and correct for the analyst's introduced bias. Also the nature of the organization and the type of incentives or rewards it offers may have

unintended, perhaps intimidating effects on the expressed preferences
and values of members of the organization. There are often major
differences between what people say that they would be willing to do
or would like to see done and what they will actually do. Thus,
values and value incoherence and inconsistency become major influences on judgment [Fischhoff, Slovic, and Lichtenstein, 1977; Sage,
1981b].

Most contemporary researchers deny the existence of unchanging
risk-taking propensity as a generalized immutable characteristic of
individuals. Also, situation determinants are much more important
than organismic characteristics such as personality. But, we have
little knowledge about the nature of these situational or environmental effects beyond the simple realization that, as the environmental similarity of decision situations decreases, generality likewise decreases [Arrow, 1974; Slovic and Lichtenstein, 1968]. The
results suggest the need to look carefully at information acquisi-
tion-and-evaluation dependencies upon the nature of various contingency task structures, for therein may lie the clue to the design of
information systems to aid in the evaluation of risk/benefit issues.
For example, it may be observed that, although there is considerable
violation of the utility axioms in decision situations in which utilities and probabilities are very different from those often experienced, the rate can fall drastically as the probability and utility
parameters are varied away from critical levels [MacCrimmon and Larson, 1979]. This suggests that utility functions defined over large
domains must be used with caution and that probabilities very near
one or very close to zero may create difficulties. Also it suggests
that descriptive cognitive style models based upon concrete operational and formal operational learning theories may be of considerable value for information system design [Sage, 1981b].

There exists, typically, a great deal of available data to aid
in the planning process. However, many information acquisition, processing, and evaluation procedures that we employ do not always tell
us, with any degree of accuracy, what we want to know. If not used
with care, they may misinform us, and leave us less well off than
we were [Kirkpatrick, Davis, and Robertson, 1976]. This may, of
course, be very harmful. There is a tendency to advocate highly
complicated system analysis models and optimization procedures without realizing that results obtained by using these procedures with
very ill-formulated issues, very poor data, and very flawed evaluation and interpretation heuristics may only represent extravagant
cost, time, and other resource expenditures with little, if any,
associated benefit. Analysis accomplished at the expense of neg-
lected issue formulation and interpretation may be worse than no
analysis at all. The simple substitution of intuitive affect may
be much less expensive and lead to better results.

There are many ways that one may proceed with useful research in risk/benefit analysis. We accept the premise that individuals generally desire the capability of making "rational" decisions, and that they desire to recognize when they do not. Further, we recognize the great value of sound and properly applied concrete operational experience in the formulation of appropriate judgments. To aid both in the making of rational judgments and in the determination of appropriate heuristics that closely approximate rational judgment, it seems that additional insights resulting from exploration of "errors" resulting from biases due to cognitive biases as well as cognitive heuristics and many other factors that impede wise judgment are research efforts with potentially significant benefit to the many application areas for risk/benefit analysis.

Among the recent descriptive theories of decision making which have been used to explain observed behavior in the Allais examples, and others, is the prospect theory of Kahneman and Tversky [1979]. Prospect theory views risky decision making as a two-phase process. The first phase involves editing the given decision problem into a simpler representation in order to make the evaluation of gambles and choice easier for the decision maker. The second phase involves assigning an overall value to each edited gamble and the subsequent choice of the gamble with the greatest value.

An important feature of prospect theory is the critical role that is attributed to an aspiration level in the analysis of risky decisions. A key operation in the editing phase is the coding by the decision maker of each of the outcomes of a gamble as being either a gain or a loss, with a gain or loss defined by the relationship of the outcome to a reference point or level of aspiration. Thus, prospect theory may allow one to satisfice as a descriptive choice-making process [Simon, 1957, 1979a, and 1979b]. According to Kahneman and Tversky, a decision maker responds to a gamble, in part, by assigning subjective values to the gains and losses associated with it. In this model, probabilities are modified by a weighting function. However, the weights enter the decision criteria linearly and are not normalized. Thus, for a given gamble, the weights need not be "coherent," that is, they do not necessarily sum to one. Furthermore, the weighting function in the prospect theory is required to have certain specific additional properties [Kahneman and Tversky, 1979].

The value function, which is the Kahneman and Tversky version of a utility function, is assumed to be concave for gains, convex for losses, and steeper for losses than for equivalent gains. Kahneman and Tversky point out that a consequence of their theory is that a change in the reference point may change the preference ordering among a set of gambles.

In models of substantively rational decision making, people
make "optimal" choices from among a set of previously identified
courses of action. The problem is typically structured by the systems analyst in the form of a tree. Alternative courses of action
or possible decisions are recognized, as is the set of consequences
or outcomes that might follow from the alternatives. Each consequence has a value or utility which is elicited from the decision
maker as well as a subjective probability of occurrence of the various outcomes. An aggregation rule is then used to aggregate facts
or probabilities, and values such as to allow prioritization of
alternatives [Millburn and Billings, 1976; Sage, 1977].

The framework that we need to use to resolve risk/benefit issues
involves, in part, a substantively rational approach. But it should
also involve process rationality. Thus the approach should be an
iterative one involving a search for a set of means-ends relationships within which a search for prescriptive solutions or means may
be undertaken [Gaul, 1977; Hammond, McClelland, and Mumpower, 1980;
Sage, 1977]. As many have pointed out, the entire decision is a
multi-step process in which problems or needs or issues must first
be identified, alternatives sought, and consequences identified, and
a choice made and implemented, followed by evaluation of the results
to see if the fundamental issue has been resolved.

Insuring that the "right" question is addressed is a priority
concern [Nisbett and Wilson, 1977]. To begin with, the definition
of most problems or issues is not given, but is the result of a
social-psychological-political process. The problem is defined
through the process of comparing the existing situation with some
desired situation or need and noting a discrepancy. The desired situation is, in essence, a model of how things should be. The decision
maker expresses needs because of a value system that is based upon
internal personal values and such things as historical data, explicit
projections, and other values of individuals [Millburn and Billings,
1976; Sage, 1977].

When the scope of an issue is defined and when the interest or
stakeholder groups involved are established, the next essential task
of the systems analyst is the determination of the goals or objectives of the decision maker and appropriate stakeholder groups.
This structural decomposition of the problem is directed toward
identifying decision and state variables and the relationships between them and explicitly structuring values by which to evaluate
outcomes.

Even after the issue is defined, alternatives and their consequences are not given but must be identified. The need for a theory
of search behavior is increasingly being emphasized [Kirkpatrick,
Davis, and Robertson, 1976; Simon, 1979a]. Values attached to

outcomes may not be known with any great certainty and may, in fact, be both labile and inchoate during the decision process [Fischhoff, Slovic, and Lictenstein, 1980]. Systems engineering support to decision making must involve much consideration of values. For example, much of the controversy concerning the SST centered around the values or utilities attached to the various outcomes of building and use of a supersonic transport by various stakeholder groups [Millburn and Billings, 1976]. The processes of defining an issue, getting it onto an agenda, and searching for alternatives--that is to say the issue formulation process--are likely to alter values. The various components of the analysis-and-interpretation steps are influenced very much by values: those of the client and those of the analyst. Analyst values do not, properly, belong in the systems process. To design processes that allow incorporation of coherently expressed stakeholder values is a major challenge.

The observed lability of human values suggests that the timing of an actual choice-making process may be crucial. Some politically astute decision makers who have recognized this often monitor the shifting of values of the decision-making body and attempt to "call the question" when the values would support their desired alternative [Millburn and Billings, 1976].

Among the important issues that must be examined during the analysis phase are those concerning the types of uncertainty characterizing different decision situations and the extent of biases due to faulty information acquisition, information processing, and other factors. Encoding of information regarding uncertainty on state variables and incorporating decision-maker attitude toward risk into the utility function are among the principal tasks of the analyst in this step of systems engineering activity.

It cannot be overemphasized that the decision-maker's judgment in any choice situation will be dependent on the body of assumptions, world view, values, professional background, methodology, and specific applicable data available [Nisbett and Wilson, 1977]. Thus the large potential for major losses, made possible by poorly structuring a real-world social problem into a quantifiable model, should not be overlooked [De Waele, 1978; Einhorn and Hogarth, 1981; Goldberg, 1976; Hershey and Schoemaker, 1980; Libby, 1976; Wallsten, 1980]. Ultimately however, judgment is also based upon procedures used in identification and acquisition of information, which is data of value in decision making, and the decision rules employed during the phases of information aggregation and evaluation. These, together with ubiquitous questions involving values, are the major areas in which contemporary advances in risk/benefit analysis need be made to insure efficacious technology-choice decisions for the betterment of humankind.

REFERENCES

Albert, J. M., Aug. 1977, Some epistemological aspects of cost bene-
    fit analysis, George Washington Law Rev., 45(5):1025-1036.
Allais, M., Oct. 1953, Le comportement de l'homme rationnel devant
    le risque: Critique des postulats et axiomes de l'ecole Ameri-
    caine, Econometrica, 21(4):503-546.
Allais, M., and Hagen, Oh., 1979, "Expected Utility Hypotheses and
    the Allais Paradox," D. Reidel, London.
Arrow, K. J., 1974, "Essays in the Theory of Risk Bearing," Elsevier,
    New York.
Bunker, J. P., Barnes, B. A., and Mosteller, F., 1977, "Costs, Risks,
    and Benefits of Surgery," Oxford, New York.
Coombs, C. H., 1975, Portfolio theory and the measurement of risk,
    in: "Human Judgment and Decision Processes," Martin F. Kaplan
    and Steven Schwartz, eds., Academic Press, New York.
Coombs, C. H., and Huang, L. C., 1976, Tests of the betweeness prop-
    erty of expected utility, J. Math. Psychol., 13:323-337.
Daniel, D. W., 1979, What influences a decision? Some results from
    a highly controlled defense game, OMEGA, Int. J. Management Sci.,
    8(4):409-419.
Detmer, R. E., Fryback, D. G., and Gassner, K., Aug. 1978, Heuristics
    and biases in medical decision-making, J. Med. Ed., 53:682-683.
De Waele, M., 1978, Managerial style and the design of decision
    aids, OMEGA, Int. J. of Management Sci., 6(1):5-13.
Driscoll, J. M., and Corpolongo, M. J., May 1980, Uncertainty esti-
    mation and the uncertainty-probability shift under information
    load, Behavioral Sci., 25:205-218.
Edwards, W., 1954, The theory of decision making, Psychol. Bull.,
    51(5):380-417.
Efstathious, J., and Rajkovic, V., June 1979, Multi-attribute
    decision-making using a fuzzy heuristic approach, IEEE Trans.
    Syst., Man, Cybern., SMC-9(6):326-333.
Einhorn, H. J., and Hogarth, R. M., 1981, Behavioral decision theory:
    Process of judgment and choice, Annual Rev. Psychol., pp. 53-88.
Fischer, G. W., July 1979, Utility models for multiple objective
    decisions: Do they accurately represent human preferences?
    Decision Sci., 10(3):451-479.
Fischhoff, B., 1975, Hindsight and foresight: The effect of out-
    come knowledge on judgment under uncertainty, J. Exp. Psychol.:
    Human Perception Performance, 1(3):288-299.
Fischhoff, B., 1977, Perceived informativeness of facts, J. Exp.
    Psychol.: Human Perception Performance, 3(2):349-358.
Fischhoff, B., 1980, For those condemned to study the past: Reflec-
    tion on historical judgment, in: "New Directions for Methodol-
    ogy of Social and Behavioral Science: Fallible Judgment in
    Behavioral Research," Richard A. Schweder, ed., Jossey Bass,
    San Francisco.

Fischhoff, B., Slovic, P., and Lichtenstein, S., 1977, Knowing with
    certainty: The appropriateness of extreme confidence, J. Exp.
    Psychol.: Human Perception Performance, 3(4):552-561.
Fischhoff, B., Slovic, P., and Lichtenstein, S., 1980, Knowing what
    you want: Measuring labile values, in: "Cognitive Processes
    in Choice and Decision Behavior," T. S. Wallsten, ed., Law-
    rence Erlbaum Associates, Hillsdale, N. J.
Fischhoff, B., Goiten, B., and Shapira, Z., 1981, The experienced
    utility or expected utility approaches, in: "Expectancy,
    Incentive and Action," N. Feather, ed., Erlbaum, Hillsdale,
    N. J.
Fishburn, P. C., 1974, Lexicographic orders, utilities, and decision
    rules, Management Sci., 20(11):1442-1471.
Gaul, M., 1977, Influence of anxiety level of evaluation of dimen-
    sional importance in risky tasks, Polish Psychol. Bull., 8(3):
    165-170.
Goldberg, L., 1976, Man versus model of man: Just how conflicting
    is that evidence? Organizational Behavior and Human Perfor-
    mance, 16(21):13-22.
Greenberg, E., Marshall, W. J., and Yawitz, J. B., June 1978, The
    technology of risk and return, Am. Econ. Rev., 68(3):241-251.
Hammond, K. R., McClelland, G. H., and Mumpower, J., 1980, "Human
    Judgment and Decision Making: Theories, Methods and Proce-
    dures," Hemisphere/Praeger, New York.
Hansson, B., 1975, The appropriateness of the expected utility
    model, Erkenntnis, 9:175-193.
Helson, H., 1965, "Adaptation-Level Theory," Harper & Row, New
    York.
Hershey, J. C., and Schoemaker, P. J., 1980, Prospect theory's
    reflection hypothesis: A critical examination, Organizational
    Behavior and Human Performance, 25:395-418.
Hogarth, R. M., June 1975, Cognitive processes and the assessment
    of subjective probability distributions, J. Amer. Stat. Assoc.,
    70:271-294.
Holsti, O. R., Sept./Oct. 1976, Cognitive process approaches to
    decisionmaking, Am. Behavioral Scientist, 20(1):11-32.
Janis, I., and Mann, L., "Decision Making," Free Press, New York.
Kahneman, D., and Tversky, A., Mar. 1979, Prospect theory: An
    analysis of decision under risk, Econometrica, 47:263-291.
Karmarkar, U. S., Feb. 1978, Subjectively weighted utility: A
    descriptive extension of the expected utility model, Organi-
    zational Behavior and Human Performance, 21(1):61-72.
Karmarkar, U. S., 1979, Subjectively weighted utility and the
    Allais Paradox, Organizational Behavior and Human Performance,
    24:67-72.
Kenney, R. L., and Raiffa, H., 1976, "Decisions with Multiple Ob-
    jectives: Preferences and Value Tradeoffs," Wiley, New
    York.

Kirkpatrick, S. A., Davis, D. F., and Robertson, R. D., 1976, The
    process of political decision making in groups: Search behav-
    ior and choice shifts, Amer. Behavioral Scientist, 20:33-64.
Koriat, A., Lictenstein, S., and Fischhoff, B., Mar. 1980, Reasons
    for confidence, J. Exp. Psychol.: Human Learning and Memory,
    6(2):107-118.
Kozielecki, J., 1974, Environment and personality as determinants
    in decision making, Polish Psychol. Bull., 5(4):3-11.
Kozielecki, J., 1975, The internal representation of risky tasks,
    Polish Psychol. Bull., 6(3):115-121.
Kozielecki, J., June 1977, Decision theory and an operational model
    of man, Polish Psychol. Bull., 8(3):127-135.
Larichev, O. I., Boichenko, V. S., Moshkovich, H. M., and Shepta-
    lova, L. P., Modeling multiattribute information processing
    strategies in a binary decision task, Organizational Behavior
    and Human Performance, 26(2):278-291.
Lewin, K. T., Dembo, T., Festinger, L., and Sears, P., 1974, Level
    of aspiration, in: "Personality and Behavior Disorders,"
    Ronald Press, New York.
Libby, R., June 1976, Man versus model of man: Some conflicting
    evidence, Organizational Behavior and Human Performance, 16:
    1-12.
Lichtenstein, S., and Slovic, P., 1971, Reversals of preferences
    between bids and choices in gambling decisions, J. Exp.
    Psychol., 89(1):46-55.
MacCrimmon, D. R., 1973, An overview of multiple objective decision-
    making, pp. 18-44, in: "Multiple Criteria Decision Making,"
    J. L. Cochrane and M. Zeleny, eds., University of South Caro-
    lina, Columbia, S. C.
MacCrimmon, K. R., and Siu, J. K., 1974, Making trade offs, Decision
    Sci., 5:680-705.
MacCrimmon, K. R., and Larson, S., 1979, Utility theory: Axioms
    versus paradoxes, in: "Expected Utility Hypothesis and the
    Allais Paradox," Maurice Allais and O. Hagen, eds., D. Reidel,
    London.
Matusewicz, C., 1975, Value or needs? How situations affect the
    course of human behavior, Polish Psychol. Bull.,6(2):101-107.
Messick, D., and McClintock, C., 1968, Motivational bases of choice
    in experimental games, J. Exp. Soc. Psychol., 4:1-25.
Milburn, T. W., and Billings, R. S., Sept./Oct. 1976, Decision-
    making perspectives from psychology, Am. Behavioral Scientist,
    20(1):111-126.
Morris, P. A., May 1974, Decision analysis expert use, Management
    Sci., 20:1233-1241.
Nisbett, R., and Wilson, T. D., May 1977, Telling more than we can
    know: Verbal reports on mental processes, Psychol. Rev.,
    84(3):231-259.

Nisbett, R., and Ross, L., 1980, "Human Inference: Strategies and
    Shortcomings of Social Judgment," Prentice Hall, Englewood
    Cliffs, N. J.
Okrent, D., ed., 1975, "Risk Benefit Methodology and Application:
    Some Papers Presented at the Engineering Foundation Workshop,
    September 22-26, 1975, Asilomar, California," UCLA-ENG-7598.
Payne, J. W., 1973, Alternative approaches to decision making
    under risk: Moments versus risk dimensions, Psychol. Bull.,
    80(6):439-453.
Pitz, Gordon F., Heerboth, J., and Sachs, N. J., Aug. 1980, Assess-
    ing the utility of multiattribute utility assessments, Organi-
    zational Behavior and Human Performance, 26(1):65-80.
Poulton, E. C., 1977, Quantitative subjective assessments are almost
    always based, sometimes completely misleading," J. Psychol.,
    68:409-425.
Roberts, H., Aug. 1978, Regulatory aspect of the food additive risk/
    benefit problem, Food Technology, 32:59-61.
Sage, A. P., 1977, "Methodology for Large-Scale Systems," McGraw-
    Hill, New York.
Sage, A. P., and White, E. B., Aug. 1980, Methodologies for risk
    and hazard assessment: A survey and status report, IEEE Trans.
    Syst., Man, Cybern., SMC-10(8):425-446.
Sage, A. P., 1981a, Systems engineering, in: "McGraw-Hill Encyclo-
    pedia of Science and Technology," McGraw-Hill, New York.
Sage. A. P., 1981b, Methodological considerations in the design of
    large scale systems engineering process, in: "Large Scale Sys-
    tems Engineering," Y. Haimes, ed., Elsevier-North Holland,
    New York.
Savage, L. J., 1954, "The Foundations of Statistics," Wiley, New
    York.
Simon, H., 1957, "Administrative Behavior," Free Press, New York.
Simon, H., 1979a, On how to decide what to do, Bell J. Econ., 10:
    494-507.
Simon, H., 1979b, "Models of Thought," Yale University Press,
    New Haven.
Sjoberg, L., Aug. 1979, Strength of belief and risk, Policy Sci.,
    11(1):39-57.
Skweder, R. A., 1980, "Fallible Judgment in Behavioral Research,"
    Jossey Bass, San Francisco.
Slovic, P., and Lichtenstein, S., Nov. 1968, Relative importance
    of probabilities and payoffs in risk taking, J. Exp. Psychol.,
    Monograph, 78, Part 2, pp. 1-18.
Slovic, P., Fischhoff, B., and Lichtenstein, S., 1977, Behavioral
    decision theory, Ann. Rev. Psychol., 28:1-39.
Sniezek, J. A., 1980, Judgments of probabilistic events: Remember-
    ing the past and predicting the future," J. Exp. Psychol.:
    Human Perceptions and Performance, 6(4):695-706.
Svenson, O., Feb. 1979, Process descriptions of decision making,
    Organizational Behavior and Human Performance, 23(1):86-112.

Swieringa, R., Gibbins, M., Larsson, L., and Sweeney, J. L., 1976, Experiments in the heuristics and human information processing, J. Accounting Res., 14:159-187.

Thaler, R., Mar. 1980, Toward a positive theory of consumer choice, J. of Econ. Behavior and Org., 1(1):39-60.

Thorngate, W., May 1980, Efficient decision heuristics, Behavioral Sci., 25(3):219-225.

Tversky, A., 1969, Intransitivity of preferences, Psychol. Rev., 76(1):31-48.

Tversky, A., and Kahneman, D., 1971, Belief in the Law of Small Numbers, Psychol. Bull., 76(2):105-110.

Tversky, A., June 1972, Elimination by aspects: A theory of choice, Psychol. Rev., 79:281-299.

Tversky, A., and Kahneman, D., 1973, Availability: A heuristic for judging frequency and probability, Cognitive Psychol., 5:207-232.

Tversky, A., and Kahneman, D., Sept. 1974, Judgment under uncertainty: Heuristics and biases, Science, 185:1124-1131.

von Neumann, J., and Morgenstern, O., 1947, "Theory of Games and Economic Behavior," Princeton University Press, Princeton, N. J.

Wallsten, T. S., 1980, "Cognitive Processes in Choice and Decision Behavior," Lawrence Erlbaum Associates, Hillsdale, N. J.

White, C. C., and Sage, A. P., June 1980, A multiple objective optimization based approach to choicemaking, IEEE Trans. Syst., Man, Cybern., SMC-10:315-326.

Wolfson, R. J., and Carroll, T. M., 1976, Ignorance, error and information in the classical theory of decision, Behav. Sci., 21:107-115.

PANEL SESSION: "THE ROLE OF GOVERNMENT IN ASSESSING THE ACCEPT-

ABILITY OF RISK AND THE EFFICACY OF SAFETY"

J. W. Bulkley,* Moderator, Theodore Schad, David Bowles,
Eleonora Sabadell, and Malcolm Simmons

*Department of Civil Engineering
 University of Michigan
 Ann Arbor, Michigan

To initiate this panel discussion, a series of four questions
was formulated prior to the meeting and these questions were included in the Conference Program. The four questions were as follows:

1. What special elements are associated with risk analysis
   and risk assessment to require governmental action?

2. How do these special elements associated with risk analysis and risk assessment differ from past requirements for
   governmental action?

3. What specific applications in the water resources field
   would benefit from risk analysis studies?

4. What requirements assure that the governmental function
   of risk analysis is properly incorporated into the decision-making process?

The entire conference has been devoted toward providing insights into the issues raised by these questions. The panelists
will also provide their own views of the role of government in the
actual implementation of risk-analysis and risk benefit to societal
problems. As moderator, let me open the discussion with certain
tentative ideas.

First, one certainly sees many opportunities for governmental
utilization of risk-analysis and risk-assessment in the important
and vital function of environmental regulation. It is, in fact,
a responsibility of government to formulate and implement necessary

263

regulations which will protect and promote public health and
safety. Often, these regulations will have an associated economic
cost. Accordingly, the government acts to undertake the trade-off
analysis for society as a whole in weighing the reduced risks resulting from certain policy choices versus the increased economic
or other costs associated with implementing the policy choice.

Secondly, it may be argued that Risk Analysis and Risk Assessment is complex and not fully understood by the public. In the
past, the role of government in providing rules and regulations to
prevent water-borne illness were clear and persuasive. A large
body of data existed which would clearly indicate the results if
water supplies were not potable. More recently, it becomes difficult if not impossible for the public to discriminate between risk
probabilities of 1/1000 versus 1/10000. Accordingly, the government has a education role as well as a decision and regulatory
role to perform. In my view, the potential exists for the tyranny
of the expert to arise in the context of risk assessment and risk
analysis. In short, how does one assure that there is a reasonable and realistic independent review of actions taken by the
experts in the field of risk analysis?

Third, many examples can be identified where government and
water resources will interact. Allowable standards for toxic
organics in water supply would be an example. Another is the
questions of stream standards for an individual state. In this
case, what benefits should be achieved at what economic cost? Is
it important for the public to be able to identify quantifiable
improvements in water quality as a result of enhanced or more
rigorous water quality standards? Additional examples would include issues of dam safety, levee failure, flood-plain occupancy
and associated risks, risks-of-failure of aging distribution and
collection systems.

Finally, Judge David L. Bazelon has provided an excellent
statement regarding the requirements for assuring that the government function of risk analysis is properly incorporated into the
decision-making process.[1]

1. There must be an explicit statement of what is uncertain.
    For example, one must not mask uncertainty by assuring
    the public that there is no risk when, in fact, the
    responsible decision-makers (experts) know that the
    answers are not available.

2. In addition to the statement of uncertainty, the decision
    process needs an explicit statement of all the problems
    attendant to the issue under examination. This informa-
    tion would include past errors or mistakes, differences

in expert opinion, and a forthright assessment of current uncertainty.

3. Finally, the end result should be a decision which is based upon calm reflection, full debate, and mature decision. It should not be a decision based upon premature closure, hasty analysis, and incomplete assessment.

Now let us consider the views of the panelists.

REFERENCES

1. David L. Bazelon, Risk and responsibility, Science, 205:277-280 (1977).

PANELIST: Theodore M. Schad
           Deputy Executive Director
           Commission on Natural Resources
           National Research Council, Washington, D.C.

Risk and safety has not generally been primary considerations in the authorization of water resources programs. Until the massive Teton Dam failure in 1976, the Congress accepted the assurances of federal officials that dams would not fail. Statements in flood control reports suggesting that levees might be overtopped by a storm expected to occur once in two hundred years or once in five hundred years fell on deaf ears in the Congress, such occurrences being perceived as being so remote that they need not be considered. Certainly hydrologists in the federal agencies were well aware of the risks, but any reservations they may have expressed in the form of language in the project authorization reports were overridden by the desire to keep the program going. Curiously, the Corps of Engineers, which is now recognized as the primary builder of large dams in the United States, resisted the use of reservoirs for flood control for over half a century after the completion of Humphrey and Abbot's report on the hydrology of the Mississippi River system because that report suggested the existence of such reservoirs would impart a false sense of security to residents of flood plains downstream from the reservoir. All such resistance had disappeared even before the Flood Control Acts of June 28, 1938 and August 18, 1941 effectively "stacked the deck" in favor of reservoirs over local protective works.

Reliance on short hydrologic records led to many an economic failure on federal reclamation projects. Witness the San Carlos reservoir in Arizona, which stood nearly empty almost half a

century because of over-design. Yet arguments that it was over-
designed now lose their cogency since the reservoir capacity was
fully used in last year's floods. Many other examples could be
cited to show that risk-benefit analysis has never been a consideration in water project authorizations.

Unfortunately, the same can also be said with respect to the
even larger authorizations for federal funding of sewage treatment
plants in the Federal Water Pollution Control Act Amendments of
1972. It is probable that that legislation, which has led to huge
expenditures for the construction of treatment plants with only
marginal improvements in water quality that provide generally in-
tangible benefits, is one of the reasons there has been more interest expressed in risk-benefit analysis in some committees of the
Congress. The Safe Drinking Water Act, calling for studies of
populations placed at risk by various levels of contaminants in
drinking water, and prescribing control technologies to protect
those at risk, suggests a rudimentary concern for risk-benefit
analysis. But my work over the years with the legislative history
of various water resources does not disclose any real interest in
the subject of risk-benefit analysis by the Congressional committees having responsibilities in water resource areas.

Unfortunately, perhaps, legislation evolves in a rather hap-
hazard way. The 1972 amendments to the Federal Water Pollution
Control Act, with their goal of elimination of discharges of pollutants into navigable waters were introduced and passed by the
Senate as a substitute for a more conventional House-passed bill.
In the conference between the two Houses of Congress which followed, redundant portions of both bills were retained, lending a
general aura of confusion to the field of Federal activity in
water pollution control which persists to this day. Again, the
primary consideration appeared to be getting the money out to the
cities, and not worrying about the risks and the benefits.

At the present time, after sitting through four days of discussions on the subject of risk-benefit analysis for water resources, and many previous meetings on the same subject, I am not
at all convinced that the water resources profession has enough
of a "handle" on the problem to develop procedures or methodologies for dealing with the subject as a part of an analysis of
water resources programs. The Water Resources Council's effort
to set down principles and standards should give us a basis on
which to start, but I hope there will be ample opportunity for
trial and error before any procedures are frozen into regulations.
What is needed right now is a set of case studies, somewhat analogous to those that were done for the draft of the Principles and
Standards before they were promulgated in 1973.

PANELIST: David S. Bowles
            Associate Professor of Civil and Environmental Engineering
            Utah Water Research Laboratory, College of Engineering
            Utah State University, Logan, Utah

First, I would like to point out that most of my experience
related to risk assessment is in the area of the risk determination
and not risk evaluation [Rowe, 1977]. In other words, I have been
primarily involved in developing and applying methods for estimating the probability of occurrence of events and the magnitude of
their consequences. My experience in the area of assessing the
acceptability of risk has been limited to the application of economic criteria such as benefit-cost analysis in which risk factors
have been incorporated in the benefits and costs [e.g., James
et al., 1979]. Therefore, my comments will reflect my background
and will perhaps provide a contrast to the federal government and
political science perspectives of my colleagues on the panel. In
fact my discussion will probably raise more questions than it will
answer on the subject of "The role of government in assessing the
acceptability of risk and the efficacy of safety."

Next, I would like to emphasize the pervasive nature of the
risks and uncertainties present in water resources planning and
management. Much of our discussion in the past few days has concentrated on catastrophic risk, such as the failure of a dam. However, there are many other types of water resources risk, some of
which are much more likely to occur than the very low probability
catastrophic risks. Investment risk is present in all water resources projects in the sense that the rate of return anticipated
at the planning stage might not be achieved because of shifts in
the demand or the prices of project services, or because of increases in operation and maintenance costs. There are also performance risks such as having insufficient water in a reservoir to
meet irrigation demands, or that a wastewater treatment plant will
malfunction, perhaps due to human error or equipment failure, and
raw or partially treated sewage will be spilt into rivers and
lakes. Less easy to quantify, but, nevertheless, important, are
the social, environmental, health, and institutional risks associated with water resources development.

Since most water resource development consists of public works
projects, all levels of government are already involved in the
planning and decision making process. For this reason, and because
one of the functions of government is to protect people from those
things against which they are unable to protect themselves [Low-
rance, 1976], government must be involved in risk evaluation. At
this conference many of the representatives from the federal
agencies responsible for water resources planning have been seeking the answer to the question, "How do we fulfill the risk and

uncertainties requirements of the new principles and standards?"
To answer their question requires addressing two issue areas:
1) what methodologies should be used for risk determination and
2) what criteria should be used for risk evaluation and how these
criteria should be established. The topic for our panel discussion relates directly to the latter issue area and, therefore, I
will confine most of my comments to the subject of risk evaluation.
Risk evaluation for water resources projects is the process of
establishing acceptable levels of risk to project users and to
society, and identifying and implementing means and measures of
risk reduction (e.g. operator training, improved equipment reliability) and risk avoidance (e.g. flood plain zoning) so that the
acceptable risk level can be achieved and maintained. The specific topics I would like to raise are the types of risk acceptability criteria, the dynamic and uncertain nature of the entire risk
assessment process, and the implications for engineering liability
of the admission that there is a residual, albeit "acceptable,"
level of risk of a structural failure.

     In his talk on Tuesday morning Dr. Rowe reviewed several cri-
teria for risk acceptance, including: "as low as practicable"
(ALAP) or "as low as reasonably achievable" (ALARA), point of marginal cost of increased risk reduction equal to the marginal benefit of the risk reduction, best practicable technology (BPT), best
available technology (BAT), and zero risk. Dr. Rowe stressed that
none of these is a generally applicable criterion. The basic question is how far should we go and how much should we pay to reduce
the various types of risks associated with water resources projects. A characteristic of recent legislation and regulations
which relate to technology, safety, and the environment, is that
universal and technology-specific solutions are sought but these
solutions are cost-ineffective in many applications. A classic
example of this in our own field is the Federal Water Pollution
Control Act Amendments of 1972 (PL 92-500) with their requirements
for BPT, BAT, and zero discharge. I hope that the process for
implementing the risk and uncertainty considerations of the principles and standards will not become frozen in a so-called universal or generally applicable approach which adds to the already
excessive cost burden of government regulation. What is needed
are separate and carefully selected criteria for each type of
water resources risk and where appropriate a flexible means for
risk evaluation that permits the level of risk acceptability to
be determined individually for each project rather than by some
arbitrary or nationwide standard of risk acceptance.

     The problem of real world risk evaluation is made all the
more complex by the uncertainties associated with the risk estimates. Quantitative risk evaluation requires estimates of probabilities of occurrence and magnitudes of consequences and in some

situations these are subject estimates which may lack the precision
or objectivity that should be desired. In the process of identifying the risks to be estimated some may be overlooked leading to the
problem of an incomplete description of the true risks. In addition, the process of risk assessment is a dynamic one for two reasons. Firstly, risk estimation is dynamic because both our project
data base and our knowledge (e.g. of environmental effects) grow
with time and therefore, over the life of a project as it proceeds
from planning into construction and operation. Secondly, the
nature of the uses of a water resources system change during its
life (e.g. sale of agricultural water rights to energy development)
and so do its potential effects (e.g. changes in land use in a
flood plain after the consequences of dam failure or flood), and
the values and risk perceptions of society. Changes in these factors effect both risk estimation and risk evaluation. In the face
of the uncertain dynamic nature of these factors it is necessary
that one or both of two strategies be adopted. The first is the
avoidance, where practical, of irreversible commitments to water
resource projects where great uncertainty exists with respect to
future conditions. The second is the adoption of a program for
risk management aimed at maintaining or improving the desirability
(optimality) of the risk acceptance decision. Examples of risk
management in this context are: operator training to maintain the
reliability of the operation of waste water treatment plants at
levels assumed at the planning stage; standby water conservation
programs of means and measures to reduce water demands and economic
losses during drought periods; frequent dam safety inspections in
an attempt to avoid increases in the risk of structural failures;
and flood plain zoning which will ensure that significant increases
in the value of land use do not take place to the point of making
a dam "unsafe" because the consequences of a failure have multiplied. The costs of such risk management programs should be borne
by the water resources project and the project should still be
economically feasible when these costs are included.

Current work at Utah State University on the probabilistic
slope stability analysis of embankment dams is demonstrating that
this procedure can be used in parallel with the conventional factor of safety approach [Sharp et al., 1981]. The probabilistic
approach has the advantage that it enables the designer to include
the variability in the factor of safety instead of looking only at
values based on conservative assumptions about material properties.
Thus alternative designs can be compared based on their probabilities of failure (i.e. probabilities of factor of safety less than
1.0) due to slope instability. But before the concept of acceptable risk can be applied to structural risk the implications for
engineering liability must be resolved. Does the acceptable risk
concept mean that a failure rate below the acceptable level would
not imply liability of the designer, assuming that negligence

could not be demonstrated? Even if this could be agreed in principle it is doubtful that it could be proven in practice because
of the very low failure rates that would likely be considered
acceptable, and because of the unique nature of water resources
structures in terms of site conditions, design, and uses. If a
legal recognition of the residual risk of structural failure is
truly unworkable what incentive is there for an engineer to utilize risk assessment in structural design?

In conclusion I am very pleased to see that the Water Resources Council has included risk and uncertainty in the new principles and standards. I am a strong believer in the potential
value of risk assessment at all phases of the water resources planning and management process. However, we must be careful not to
expect too much from risk assessment at this early stage. There
is a need for education of water resource planners and decision
makers in the application of risk assessment to water resources
planning and management. Procedures for using experts to give
good probability estimates must be developed. There is a need for
continued research to develop practical risk assessment procedures.
The water resources planning and management process must be carefully examined to determine where and how to include risk evaluation procedures in an hierarchical manner that will keep the volume
of information manageable and the tradeoffs clear. If the area is
"oversold" at this early stage there is a danger that it will not
measure up to our expectations and that it will be abandoned without a fair trial. What must be avoided is the adoption of arbitrary procedures and risk acceptance criteria which may lead to
cost-ineffective solutions in many situations. In summary, I
believe that risk assessment will be a valuable addition to water
resources planning and management if it is implemented at a reasonable pace and with the necessary education of both planners and
decision makers.

REFERENCES

James, L. D., Bowles, D. S., James, W. R., and Canfield, R. V.,
    1979, "Estimation of Water Surface Elevation Probabilities
    and Associated Damages for the Great Salt Lake," Report
    UWRL/P-79/03, Utah Water Research Laboratory, College of
    Engineering, Utah State University, Logan, Utah.
Lowrance, W. W., 1976, "Of Acceptable Risk: Science and the
    Determination of Safety," W. Kaufmann.
Rowe, W. D., 1977, "An Anatomy of Risk," Wiley, New York.
Sharp, K., Anderson, L. R., Bowles, D. S., and Canfield, R. V.,
    1981, A model for assessing slope reliability, Proc. 1981
    Annual Meeting Transportation Res. Board.

PANELIST: Eleonora Sabadell
            Program Director, Decertification Assessment in the U.S.
            Bureau of Land Management
            U.S. Department of the Interior
            Washington, D.C.

I would like to expand on what was discussed during this morning's session, about the type of information which results from
risk/benefit analysis and the type of information required by politicians and decision-makers.

It is not only the format in which the information should be
made available to the politician and decision-maker but also the
scope of the analysis to be made, and the range of possible alternatives offered, that may be substantially different from a risk/
benefit analysis performed for (as an example) the construction of
an engineering project, e.g., a dam. Allow me to illustrate this
point.

In the case of the engineering project the probability of
failure must be minimized toward zero. In the case of a politician
taking a stance or making a decision, this person may be willing to
accept a much higher level of risk of initially failing so as to
gain an eventual or a delayed success.

Where in the building of a dam the objectives are definite
and the outcomes, risks and benefits are quantifiable. However,
in the development of a policy, the objectives to be taken into
account may evolve, change, and even conflict (fisherman vs. power
plant operator on instream flow and water quality matters), and
the outcomes, risks and benefits may be given different values by
the different people involved.

Finally, the range of alternatives and trade-offs a decision-
maker should consider is in general much broader than the alternatives and trade-offs acceptable to the builder of the dam in our
example.

In summary, I believe the risk analysis community should
recognize the diversity of "users" and their requirements, thus
tailoring their assistance, specially to the decision-makers, in
such ways as to satisfy those various and different needs. This
signifies the development and application of new and more creative,
flexible approaches to risk analysis, where not only the physical
systems are taken into account but where the "human factor" is considered as well.

PANELIST: Malcolm Simmons*
            Analyst, Environment and Natural Resources Policy Division
            Congressional Research Service, The Library of Congress
            Washington, D.C.

Process v. Method and Risk Perception

One of the points brought out in the conference was process
versus method, where process was the procedural mechanism used in
making a decision, and method the techniques or analytical devices
used. Another topic of discussion in the conference on risk assessment was risk <u>perception</u>.   I offer that the two topics should be
integrated into a common theme.

Mr. James W. Spensley elucidated the Congressional perspective on process, pointing out how concerned interest groups should
participate from the outset. The failure to consider the viewpoints
of concerned interest groups from the outset could lead to problems
of public acceptance at a later point in time. Even though the
opinions of the different interest groups might differ at the outset, discussion of the differences and understanding the reason for
these differences is likely to result in a meeting of minds and a
mutually acceptable course of action.

The discussions of method reviewed various models where the
analysts attempted to assess the risk of alternative courses of
action. Nowhere in the conference, however, has there been a discussion of the possibility of including an analysis of the risk
<u>perception</u> of the concerned interest groups as part of the method.
Inclusion of this kind of information along with the risk assessments of alternative courses of action would provide a useful tool
for the policy maker and make him/her fully aware of the differing
opinions on various courses of action.

With this kind of information in hand, policy makers could
work with interest groups in opposition to a particular course of
action to find out if there is validity to the difference of opinion. If the opposing viewpoints are based on misapprehension of
the facts, the true facts could be presented and the differences
resolved. On the other hand, the opposition could be based on
facts not considered in the policy-making process theretofore,
and the opposing viewpoints of certain concerned interest groups
could influence the opinions of other interest groups on a particular course of action. Future policy deicions need the broad based
support of concerned interest groups, and integration of risk

---

*The views expressed in this commentary are those of the author and
do not necessarily reflect the views or positions of the Congressional Research Service.

perception into method or analytical technique of a decision could be of great value to the process employed by the policy makers in an attempt to find an acceptable course of action.

How Confident is Confident?

The presentations of two conference participants led me to question the confidence we should have in many technical judgments. Dr. Paul Slovic and Dr. David Campbell both used the 90 percent confidence interval to illustrate their point on this matter. A 90 percent confidence interval is a range in which we are sure that the correct answer lies 90 percent of the time. Thus I might give a 90 percent confidence interval of 200 to 300 years for the time it takes the planet Pluto to revolve around the sun in earth years (The actual answer is 248.42 years). If ten questions of this type were asked, the correct answer should fall in the range I selected for my 90 percent confidence interval for nine of the ten questions.

Dr. Campbell presented ten questions of this nature to the conference participants and asked them to give a 90 percent confidence interval. Much to the surprise of us all, the number of "correct" confidence ranges varied from one to six for the completed questionnaires. The average should have been nine correct confidence ranges. Dr. Slovic verified that on similar tests he had given, respondents actually gave a 30-50 percent confidence interval when they thought they were giving a 90 percent confidence interval. He added that when technical experts were asked questions in their own field, the actual confidence level of the response was far less than the perceived confidence level.

Such provocative tests lead to the conclusion that we may be far too confident in our technical judgments. If we apply this conclusion to decision making, we should allow a larger margin of error for the technical judgments made in risk assessment. By the same token, we should consider seriously non-technical opinion in the decision-making process.

Value of Life

I have always been disturbed by the risk assessment method of assigning a dollar value to a life, but will not go through the litany of reasons because they have been documented by many others before me. My feeling is that if risk assessors use this technique, there is a strong possibility that many people will not accept the concept of risk assessment because of the technique used.

Mr. John Graham's presentation relieved me of much of this fear. He discussed a different but related approach to risk assessment--evaluating the cost per life saved. He reviewed all the current studies employing this technique. In terms of public acceptance of the concept of risk assessment, the cost-per-life-saved approach seems far more acceptable than assigning a dollar value to life.

How Does Risk Assessment Fit Into a Highly Political Situation?

There are numerous examples of a highly political issue where risk assessment was not used at the outset of the decision-making process. One example is the controversy in California over water allocation in the Central Valley. The northern parts of the Valley are trying to safeguard as much water as possible for future use, while interests in the southern part of the Valley and the Sacramento-San Joaquin delta are asking for more water from the north for immediate use.

In such a highly emotional and political context, the mention of risk assessment seems almost out of place. Risk assessment--when used properly--is a rational process which perhaps cannot survive in a non-rational situation. Hopefully risk assessment can be used at the outset of the policy formulation process before the non-rational stage is reached.

* * * * *

CONCLUDING OBSERVATIONS

The panelists have all provided useful and stimulating insights regarding the role of government in risk analysis activities. One theme which is developed by the panelists is one of cautious optimism regarding the utilization by government of the techniques of risk-analysis and risk-assessment. It is important to provide flexibility in the application of these concepts to actual situations and one should not prescribe a rigid set of guidelines at this stage. Rather, one should be providing for an analysis environment which stimulates the development of critical analysis and insights which risk-analysis has a potential for contribution to decision processes. Clearly research is needed to facilitate the application of the technique to problem situations in order to identify the limitations as well as the strengths of the approach.

The panelists concur that the principles and standards prepared by the Water Resources Council provide a starting basis on which to develop appropriate risk-analysis techniques. At the same time, it is recognized that learning must take place and that the actual risk-assessment process may need to be done on a case-by-case basis until sufficient experience has been obtained to formulate nationwide detailed risk analysis procedures. It is pointed out that the process of risk assessment itself may lead to improved operation of facilities in order to <u>minimize</u> to the degree feasible, the probability of a class of failures being observed.

The panel also raised the important question of both format and content of a risk assessment performed for certain political leaders and decision makers. The form and type of information may be different for these individuals in comparison with an analysis on risk performed by a design engineer.

A very interesting idea raised by the panel is the question of risk perception of interest groups interacting to influence a policy-outcome. This idea is an extension of the concept of individual and group utilities serving as a motivation for action. Here, the motivation for action may reflect the individual's <u>perception</u> of risk. This perception of risk may have been little, if any, relation to the actual risk. Full consideration of perceived risks could certainly contribute to Judge Bazelon's challenge-- namely--that the end result should be a decision based upon calm reflection, full debate, and mature decision. It should <u>not</u> be a decision based upon premature closure, hasty analysis, and incomplete assessment.

SUMMARY, CONCLUSIONS, AND RECOMMENDATIONS

The following are some of the major and important issues/aspects associated with risk assessment (R.A.) compiled from a questionnaire filled by the participants on the fourth day of the meeting:

I.   Important Issues/Aspects/Elements

1) Framing the proper questions for application of risk/benefit analysis.

2) Making explicit any exercise of value judgment.

3) Translating the results into layman's terms for use by the decision makers.

4) Quality control of risk/benefit analysis.

5) Ground-water contamination as a candidate for the successful use of risk/benefit analysis.

6) Probability of failure to major structures.

7) Confidence bonds for project benefits, project costs--capital and operational, and probability of specific environmental impacts.

8) Understanding the data requirements for risk assessment.

9) Analysts'/decision makers' understanding of public perceptions, needs, and preferences concerning the reliability/risk associated with projects.

10) Defining clearly, for each situation, the objective one
    seeks; i.e., is it the need to evaluate extreme events
    (via risk/benefit analysis) or to establish probable
    outcome (risk and uncertainty)?

11) Need to implement the idea of risk and uncertainty to pro-
    ject formulation in order to establish the habit of think-
    ing in terms of probable outcomes.

12) Risk estimation techniques.

13) Risk/benefit trade-off decision process.

14) Need for a clearing house that collects information on
    events subject to risk/benefit analysis and makes data
    or analysis available to practitioners.

15) Need to very clearly differentiate between "risk" and the
    using of certain data--which should be referred to in
    terms of its degree of reliability.

16) Need to convey the reliability of the data used to the
    planners, the public, the decision-makers, etc.

17) Social-institutional barriers to implementation of
    results.

18) Importance of simplicity for the effective utilization
    of risk/benefit techniques.

19) The persistence of value judgment as the dominating
    factor in water resources decision making.

20) The equal importance of the process and the methodology
    in risk/benefit analysis.

21) Need for risks to be explicitly identified and addressed,
    rather than hidden in subjective judgments.

22) Need for improved communication among water resource
    analysts as well as between analysts and layman decision
    makers.

23) Lack of objectivity of ultimate decision makers, who are
    concerned mainly with getting the most for themselves or
    their constituents.

24) Catastrophic risk.

25) Development of <u>practical</u>, simple methodologies for risk/benefit analysis.

26) Relating risk assessment to safety factors in engineering design.

27) Developing practical interactive multiobjective methods which display the alternatives in a way that is easily perceived by the decision makers.

28) "Optimizing" the operating and planning of water systems wherever reasonable estimates of probabilities can be made.

29) The role that risk/benefit analysis could potentially serve in the communication of the planning process.

II. Important New Ideas/Concepts

1) The use of sensitivity analysis to integrate the uncertainties in risk/benefit analysis.

2) The different perceptions of the use of risk/benefit analysis and the lack of thought given to "the process" versus the methodologies.

3) Lack of consensus concerning the definitions and procedures that are desirable in risk/benefit analysis.

4) Current status of risk/benefit use in water resources (mainly supply), and the importance of political issues in water resources.

5) Lack of standardization of risk/benefit analysis--thereby making it difficult for the practitioner to accept.

6) Exposure to various approaches and views.

7) Stratification of risk/benefit analysis by purpose or other principles.

8) A better understanding of the state-of-the-art of implementing risk/benefit analysis and of its <u>complexity</u>.

9) A better understanding of the degree of difference in perceived and actual risk.

10) The infancy of risk/benefit methods, and the potential for development.

11) The identification of technical leaders in the field.

12) The imperative need for prominent display of the uncertainty band in reported results.

13) Some understanding of:

   a) people's perception of risk.

   b) approaches to risk/benefit analysis and recognition that a single approach is not available nor desirable.

   c) probabilities of rare events and difficulties of estimating them.

14) The high level of confusion about risk and uncertainty in the context of decision analysis.

15) The low level of progress to date and the rudimentary state-of-the-art of risk/benefit analysis.

16) The idea of relative probabilities proposed by Bill Rowe makes sense, but only as a screening procedure to eliminate some of the least-considered outcomes.

17) As professionals we still are not willing to discuss openly the risks of structural failure. In some respects we are not able to do so, e.g., foundation failure from undetected weakness.

18) The working of Congress; background of the principles and standards.

19) Lack of confidence displayed by speakers in risk/benefit analysis today--great deal of "negative thinking."

20) North's applications of simple probability distribution function.

21) Rowe's conception of risk.

22) Slovic's concepts on perception and risk statements.

23) Hall and Haimes' concepts on risk-benefit analysis in a multiobjective framework.

24) Campbell's 3×3 matrix approach.

25) Howard's concept of micromorts.

26) Imperative that a consistent terminology and analytical process be adopted.

27) Adjusting uncertainty and risk assessments to suit the type and quality of information available.

III. Issues Needing Further Study

1) Consensus on definitions, uses, and processes connected with successful application of risk/benefit analysis.

2) Identification and discussion of successful risk/benefit analysis to real-world decisions.

3) Rare events.

4) Contamination of drinking water--mainly ground-water sources (aquifers).

5) Better measures of benefits.

6) Methodologies for reliable estimation of risks.

7) "Acceptable" level of risks for different types and scope of projects.

8) Drawing in the public in order to develop <u>confidence</u> in the results.

9) Need for work on probability of environmental impacts.

10) Methodologies which reveal social preferences as they pertain to risk taking.

11) Understanding long-term risks that are not catastrophic rare results--e.g., long-term elimination of pristine environments in urban areas, etc.

12) How to educate users, i.e., planners and managers, to think in terms of probable outcome?

13) How to assess, accurately, the expected demands for water resources outputs?

14) Identification of critical issues in risk/benefit analysis.

15) Methodologies in risk/benefit analysis.

16) Dealing with uncertainties.

17) Comparative study of risk/benefit methodologies, particularly in connection with the formulation and analysis of water resources projects.

18) Methodologies for presenting "reliability" information in project reports, including simplified methods of display.

19) Extension and refinement of multivariable decision-making analysis procedures for use in the field of water resources where the problems are to be analyzed in a multiobjective framework.

20) Develop and test educational material to help people understand risk trade-offs and actual risks.

21) Develop more flexible planning methods to cope with uncertainty.

22) Risk/benefit analysis doesn't replace decision making.

23) Improve coordination of the entire decision-making process.

24) Need for post-"project" or post-planning evaluations of the contributions, impacts, and costs of risk/benefit analysis.

25) Need to identify evaluation methods for the assessment and presentation of risk/benefit analysis in a timely fashion.

Engineering Foundation Conference

on

Risk/Benefit Analysis in Water
Resources Planning and Management

September 21 - 26, 1980

PARTICIPANTS LIST

Mr. Don Ator
Manager Econ & Planning
Gulf So. Res. Inst.
P.O.Box 14787
Baton Rouge, LA 70989

Mr. Paul E. Barker
Sr. Proj. Manager
Espey, Huston & Assoc.
4022 West Alabama
Houston, TX 77027

Mr. William Baron
Clemson University
Civil Engineering Dept.
Clemson, SC

Mr. Denis Bechard
Mathematician
Hydro Quebec
75 West Dorchester
Montreal, CANADA

Mr. John Blair
Gingery Assoc.
2840 So. Vallejo St.
Englewood, CO 80110

Dr. Gib V. Bogle
Res. Fellow Engineering
California Inst of Tech
Env. Quality Lab.
Pasadena, CA 91125

Dr. David S. Bowles
Utah State Univ.
Utah Water Res. Lab.
UMC 82
Logan, UT 84321

Mr. Johathan Bulkley
University of Michigan
Civil Engineering Dept.
108 Engineering IA
Ann Arbor, MI 48109

Dr. Nathan Buras
Visiting Professor
Stanford University
Operations Department
Stanford, CA 94305

Mr. Michael W. Burnham
US Army Corps of Engin.
Hydrolic Eng. Ctr.
609 Second Street
Davis, CA   95616

Dr. David C. Campbell
Economics-Policy Office
US Water Resources Coun.
2120 L Street N.W.
Washington, D.C.   20015

Mr. Robert Chang
Ministry of Natl. Resources
740 York Mills Rd.
Toronto, Ontario
M3B 1W7, CANADA

Mr. Robert Childs
Corps of Engineers
650 Capital Mall
Att: SPKOP-T
Sacramento, CA   95814

Mr. Gary Cobb
Water Res. & Tech.
Dept. of the Interior
18 & C Street NW Rm 4412
Washington, D.C.   20240

Dr. Jared Cohon
The Johns Hopkins Univ.
Dept. of Geo. & Env. Eng.
Baltimore, MD   21218

Dr. Sandford S. Cole
Director of Conferences
Engineering Foundation
345 E 47th Street
New York, New York   10017

Mr. Vincent Covello
Chief Program Mgr.
National Science Foundation
Div. of Policy Res. & Anal.
Washington, D.C.   20550

Mr. Ronald R. De Bruin
Chief Plans Formulation
Corps Engineers Southwest
1200 Main Tower Bldg. 520
Dallas, TX   75080

Mr. Henry L. DeGraff
US Dept. of Commerce
14th & Constitution Ave.
Rm. 589 8C
Washington, D.C.   20230

Mr. Lloyd A. Duscha
Off. Chief of Engineers
DAEN-CWE
20 Mass. Ave. N.W.
Washington, D.C.   20314

Dr. Leo Eisel
U.S. Water Res. Council
2120 L Street N.W.
Suite 800
Washington, D.C.   20037

Dr. Donald J. Epp
Asst. Dir, Land & Water
Pennsylvania State Univ.
110 Land & Water Building
University Pk, PA   16802

Dr. Ambrose Goicoechea
Corps of Engineers
Inst. for Water Resources
Kingman Building
Fort Belvoir, VA   22060

Dr. Alvin S. Goodman
Professor
NY Polytechnic Inst.
333 Jay Street
Brooklyn, NY   11201

Mr. John Graham
Nat'l Academy of Sciences
CORADM, Rm. JH-818
2101 Constitution Ave. NW
Washington, D.C.   20418

Dr. Yacov Y. Haimes
Case Western Reserve University
Systems Engineering Dept.
10900 Euclid Avenue
Cleveland, Ohio 44106

Dr. Warren A. Hall
Colorado State University
Civil Engineering Dept.
Fort Collins, CO 80521

Mr. William Hammond
Sr. Engineer
Dept. of Water Resources
P.O.Box 388
Sacramento, CA 95802

Mr. Joseph S. Haugh
U.S. Dept. of Agriculture
Soil Conservation Serv.
P.O.Box 2890
Washington, D.C. 20013

Dr. James P. Heaney
Professor
Univ. of Florida
A. P. Black Hall
Gainesville, FL 32611

Mr. Joe C. Hise
US Corps of Eng. Port Dist.
Economics Sect. EN-PLS
P.O.Box 2946
Portland, OR 97208

Dr. Daniel H. Hoggan
Professor
Utah State University
Utah Water Research Lab.
Logan, UT 84322

Mr. Ronald Howard
Stanford University
Terman Eng. Center
Dept. of Eng. Econ. Syst.
Stanford CA 94305

Mr. Steven Kasower
Staff Economist
Dept. of Water Resources
P.O.Box 388
Sacramento, CA 95802

Mr. Alonzo D. Knapp
Water & Power Resources Serv.
Bldg. 20, Denver Fed. Ctr.
Box 25247
Denver, CO 80225

Mr. Harvey Roy Kurzon
Corps of Engineers
US Army Eng. Div NCDPD
536 S Clark Street
Chicago, IL 60605

Mr. Steve Montfort
Corps of Engineers NPO
220 NW 8th Street
P.O.Box 2870
Portland, OR 97208

Dr. W. Scott Nainis
Arthur D. Little, Inc.
Operations Research Section
Acorn Park
Cambridge, MA

Dr. Ronald North
University of Georgia
Institute of Natural Resources
Athens, GA 20601

Dr. David Okrent
Univ. of California
Engineering & Appl. Sci.
5532 Boelter Hall
Los Angeles, CA 90024

Mr. W. Glenn O'Neal
Tennessee Valley Authority
Flood Haz. Anal. Branch
Div. of Water Resources
Knoxville, TN 37902

Dr. Richard Neal Palmer
Washington University
Dept. of Civil Engineering
FX-10
Seattle, WA 98195

Mr. Steven W. Pedersen
MN Water Planning Board
600 American Center Bldg.
150 Kellogg Building
St. Paul, MN 55101

Mr. David Robb
Director, Compr. Planning
St. Lawrence Seaway/USDOT
800 Independence Ave. SW
Washington, D.C. 20591

Dr. William Rowe
American University
Inst. for Risk Analysis
Washington, D.C. 20016

Dr. Eleanora Sabadell
Dept. of the Interior
Bureau of Land Management
200-A
Washington, D.C. 20240

Mr. Theodore M. Schad
Nat'l Academy of Science
Commission on Natural Res.
Room JH 828
Washington, D.C. 20418

Dr. David F. Schuy
Eng. & Res. Center
Water & Power Res. Svs.
CODE D-735 P.O.Box 25007
Lakewood, CO 80225

Dr. Malcolm Simmons, Analyst
Library of Congress
Congressional Res. Serv.
10 First Street SE
Washington, D.C. 20540

Mr. Paul Slovic
Decision Research
1201 Oak Street
Eugene, OR 97401

Dr. Soroosh Sorooshian
Asst. Professor
Systems Engineering Dept.
Case Western Reserve Univ.
Cleveland, Ohio 44106

Mr. James W. Spensley
Sci. & Tech. Committee
U.S. House of Representatives
Rayburn Office Building
Washington, D.C. 20515

Mr. A. M. Stelle, Director
Metro Water Dist. of
 Southern California
24355 Little Valley Road
Calabasas, CA 91302

Mr. Kyosti Tarvainen
Case Western Reserve Univ.
Systems Engineering Dept.
Cleveland, Ohio 44106

Mr. Peter A. Tennant
Ohio River Valley Water
Sanitation Commission
414 Walnut Street
Cincinnati, OH 45150

Dr. Bi-Huei Wang
Associate
Hazra Engineering Co.
150 S. Wacker Drive
Chicago, IL 60606

Mr. Daniel R. Yribar
Hydraulic Engineer
Water & Power Res. Svcs.
Box 043
550 W. Fort Street
Boise, ID 83724

INDEX

Abortion, 196, 198
Acid rains, 2, 201
Acrylonitrile, 237
Administrative Conference, 24
Agricultural Research Service, 24
Air pollution, 236-237
Alcohol, 205, 206, 208, 209, 212
Alcohol, Drug Abuse, and Mental
    Health Administration, 24
American Association for the
    Advancement of Science,
    175
Antibiotics, 205, 208, 212
Aquifers, contamination of, 2,
    281
Arsenic, 54, 55, 56, 237
Asthma, 197, 198
Aviation, 144, 204, 205, 208, 209

Bayesian statistical models, 10,
    11, 63, 150-154
Benefit-cost analysis, 8-10,
    233-240
Benefit-cost-risk analysis,
    135-147
Benefit-risk analysis, 98
Benzene emissions, 237
Bernoulli distribution, 73, 75-79
Bicycles, 205, 208, 209
Botulism, 196, 197, 198
Brown's Ferry Nuclear Plant, 201
Bureau of Land Management (BLM),
    23, 224, 229
Bureau of Reclamation, 164, 165,
    229

Cancer, 197, 198, 239

Carcinogens, 237, 240
Center for Disease Control (CDC),
    24
Chance-constrained programming
    model, 5, 6, 10
Chemical fertilizers, 221,
    224-225
Chromium, 55, 56
Clean Air Act, 236, 237
Clean Water Act, 56
Clothing, 236
Coal mining, 221, 226, 237
Cognitive heuristics, 246, 248,
    251, 255
Committee on Science and
    Technology, 176
Congressional Budget Office, 24
Construction, 205, 208
Consumer Product Safety
    Commission (CPSC), 23,
    177, 236, 237, 238, 239
Contraceptives, 204, 205, 206,
    208
Corps of Engineers, 7, 8, 128,
    130, 164, 165, 265
Cost-benefit analysis, 99
Cost/Benefit/Risk analysis
    (C/B/R), 149-156
Cost-effective analysis, 234-240
Council on Environmental
    Quality, 24
Cyclamates, 202

Decision making
  benefit-cost-risk analysis in,
    135-145
  cognitive biases in, 246,
    248-251, 253, 255

287

Decision making (continued)
   definition, 245
   incorporation of risk and
      uncertainty in, 85-86
   theories, 246-251, 255-257
   value judgments in, 85-86
Department of Defense, 24
Department of Health and Human
      Services (HHS), 177,
      236, 237, 238, 239
Department of Housing and Urban
      Development (HUD), 45,
      46
Diabetes, 196, 197, 198
Dicyclopentadiene (DCPD), 154
Dissolved solids, 223, 225
Drowning, 144

Earthquakes, 82, 83, 203
Economic Regulatory Administra-
      tion, 23
Electric power
   as a risk factor, 203, 204,
      205, 206, 212, 213
   fatalities from, 144, 208, 209
Electrocution, 197, 203
Emphysema, 198
Environmental Protection Agency
      (EPA), 23, 53, 54, 56,
      57, 72, 236, 237, 238,
      239
   risk analysis research at,
      28-29

Falls, 144
Fault tree, 98-99, 193
Federal Aviation Administration,
      23
Federal Emergency Management
      Agency (FEMA), 45
Federal Energy Regulatory
      Commission, 23
Federal Highway Administration,
      23
Federal Maritime Commission, 23
Federal Railroad Administration,
      23
Federal Water Pollution Control
      Act Amendments, 224,
      266, 268

Fire, 144, 198, 205, 208, 212,
      236, 237
Firearms, 144
Fish and Wildlife Service, 23,
      229
Flood, 127, 266, 167, 197, 198
Flood Control Act, 265
Flood plain, 41-52, 201-203,
      264, 265, 269
Fluorides, 225
Food and Drug Administration, 23
Food coloring, 204, 205, 208
Food preservatives, 204, 205,
      208
Food Safety and Quality Service,
      23
Football, 204, 205, 208
Forest Service, 23, 229
Fossil fuel, 2, 204

General Accounting Office (GAO),
      24, 50, 94
Genetic screening, 236
Gleason's Clinical Toxicity
      Index, 55
Ground water, 219-220
   pollution of, 154, 194, 225,
      277, 281

Handguns, 204, 205, 206, 208,
      209, 212
Hazard Index, 55
Hazard management, 25-26
Heart disease, 195, 197, 236,
      237
Highway safety, 236, 237
Home appliances, 204, 205, 206,
      208, 209, 212
Homicides, 196, 197, 198, 200
House Committee on Science and
      Technology, 22, 90
Hunting, 205, 206, 208, 212

Library of Congress Research
      Service, 24
Lifesaving programs, 233-240
Lightning, 198
Linear Systems Yield Determina-
      tion (L-SYD) Model, 131
Love canal, 14

Machinery, 144
Masked events, 84-85
Mass balance analysis, 34, 35,
        36, 37
Materials Transportation Bureau,
        23
Meteorite, 194
Mine Safety and Health Administra-
        tion, 23
Mississippi River System, 265
"Monte Carlo" simulation
        technique, 35, 37,
        158-161, 170, 172
Motorcycles, 204-206, 208, 209
Motor vehicle, fatalities from, 144,
        203, 205, 206, 208, 209
Motor vehicle accidents, 195,
        197, 198
Motor Vehicle Safety Act, 237
Mountain climbing, 204, 205, 208,
        209, 212
Multiobjective analyses, 103-118

National Academy of Sciences,
        24, 28
National Bureau of Standards, 24
National Fire Prevention and
        Control Administration,
        24
National Flood Insurance Act, 44
National Flood Insurance
        Program, 41-52
National Highway Traffic Safety
        Administration (NHTSA),
        23, 236, 237, 238, 239
National Institute for Occupa-
        tional Safety and Health,
        24
National Institutes of Health, 24
National Oceanic and Atmospheric
        Administration, 23
National Park Service, 23, 229
National Research Council, 23, 56
National Science Foundation
    risk-analysis research at,
        22-28
National Science Council, 177
National Security Council, 24
National Transportation Safety
        Board, 24
Nitrates, 177, 225

Normative decision theory, 247
Nuclear power
    fatality estimates, 208, 209
    perceived risk from, 205, 206,
        207, 210, 211, 213
Nuclear power plant, meltdown,
        73, 79
Nuclear reactor, accidents,
        80, 81
Nuclear Regulatory Commission
        (NRC)
    reactor safety studies, 63,
        75, 201
    risk analysis research, 29
Nuclear weapons, 80, 194

Occupational Safety and Health
        Administration (OSHA),
        23, 177, 236, 237, 239
Office of Management and Budget,
        24
Office of Naval Research, 24
Office of Science and Technology
        Policy (OSTP), 176, 177
Office of Technology Assessment,
        24
Office of Water Research and
        Technology (OWRT), 1
Ordinary event, 73

Pertussis vaccine, 237
Pesticides, 204, 205, 208, 221,
        224-225
Poisoning, 144
Poisson distribution, 7, 73, 79
Police work, 204, 205, 208, 212
Power mowers, 204, 205, 206
Pregnancy, 196, 197, 198
Principles, Standards and
        Procedures of the Water
        Resources Council, 6, 7,
        101, 157-161, 188, 191,
        192, 266, 274
Probability
    distribution, 170-173
    estimation of, 74-75
Public Health Service, 23

Radioactive waste disposal,
        80, 82
Railroads, 205, 206, 208, 209

Railway accidents, 144
Rare event, 73-75, 82-84
Recombinant DNA, 193
Regulatory Council, 24
Reservoir, 128, 130, 221, 223,
        265
Resource Conservation and
        Recovery Act, 56
Resources availability, 33-36
Risk
    absolute, 78, 79-81, 83
    classification of mechanisms,
        97
    definition, 60-62, 65, 101,
        280
    procedures for analysis of,
        164-165
    relative, 78-83
    sources of, 101
Risk acceptance, 91, 96, 114,
        268, 269, 271
Risk analysis research
    factors in, 90
    federally supported,
        22-29
    probability distributions,
        165-170
    relevance of, in decision
        making, 14-17
    role of government in implemen-
        tation of, 263-265
Risk assessment
    biases in, 195-202, 251
    components of, 62, 65, 90
    concepts in, 102, 279-281
    confidence in, 273
    importance in decision making,
        25, 175-179
    issues in, 277-279, 281-282
    methodological problems in,
        59-60
    methodologies, 97-99, 102-107
    models, 95-97
    overconfidence in, 200-201
    research needs in, 253-257,
        281-282
    skepticism to models of, 93-95
Risk aversion coefficient, 250

Risk/benefit analysis
    considerations in, 38-40
    methodology in, 192
    multiobjective analysis in,
        124-133
    process, 192
    uncertainty aspects in, 101,
        103
Risk determination
    components of, 62, 64
    factors in, 90
    methodologies in, 267, 268,
        269
Risk estimation
    objectivity and subjectivity
        in, 63-73
Risk evaluation, 267, 268-269,
        270
    components of, 62, 64, 80
    factors in, 90
Risk perception, 195, 272-273,
        275
    considerations in, 204-216

Saccharin, 237
Safe Drinking Water Act, 266
Salinity, 221, 223-224
Sedimentation, 5, 221
Sensitivity analysis, 107-118
Skiing, 205, 206, 208, 212
Small pox, 195, 197, 198
Smoke detectors, 236, 237
Smoking, 205, 206, 208, 209, 212
Soil Conservation Service, 7,
        164, 165
Spearman rank correlation
        coefficient, 239
Spray cans, 204, 205, 208
Sting, 198
Stomach cancer, 196, 197, 198
Stroke, 196, 197, 198
Structural failure, 7-9
Subjective expected utility
        (SEU) theory, 246-251
Suffocation, 144
Suicide, 200
Surface water
    contamination of, 2, 54-57,
        220-225

Surgery, 204, 205, 206, 208, 209
Surrogate Worth Tradeoff (SWT)
        method, 103-118
Swimming, 204, 205, 208, 209

Technology Assessment and Risk
        Analysis (TARA) Group,
        22-28
Tennessee Valley Authority, 7
Teton Dam, 7, 201, 265
Three Mile Island, 14, 86, 89,
        90, 99
Tornadoes, 195, 196, 197, 198
Toxic and Hazardous Materials
        Agency, 150
Toxic materials, containment,
        149-150
Toxic Substances Control Act, 150
Trihalomethanes, 2, 54
Tuberculosis, 197, 198

Urban Mass Transportation
        Administration, 23
U.S. Coast Guard, 23

Vaccinations, 197, 198, 204, 205,
        206, 208, 209, 212
Value judgment, 60, 85-86, 102, 271
Vandalism, 199
Vinyl chloride, 237
Volcanic eruption, 194

Washington Suburban Sanitary
        Commission, 128, 129
Water and Power Resources
        Service, 7, 229
Water-energy relationship,
        226-227
Water projects, on Indian
        reservations, 227-228
Water quality criteria, 55
Water Resources Council, 6, 7,
        10, 11, 101, 108
    risk analysis in principles,
        standards, and procedures
        of, 157-161
Water resources planning
    consideration of structural
        failure in, 7-9
    risk and uncertainty in, 101
    risk assessment in, 267-270
    relevance of statistical
        approaches in, 32-33
    role of government in, 181-189
Water rights, 184, 189, 227,
        228-230
Water supply
    effect on economic variables,
        221-222
Water transport, 144

X-rays, 201, 204, 205, 208, 209,
        212, 213